Contractor's Pricing Guide

Framing & Rough Carpentry 1996

Senior Editor
Robert W. Mewis

Contributing Editors
Thomas J. Akins
John H. Chiang, PE
John H. Ferguson, PE
Mark H. Kaplan, Jr.
Melville J. Mossman, PE
John J. Moylan
Jeannene D. Murphy
Jesse R. Page
Michael J. Regan
Kornelis Smit
William R. Tennyson, II
Phillip R. Waier, PE
James N. Wills
Rory Woolsey

Manager, Engineering Operations
Patricia L. Jackson, PE

Vice President, Editorial Operations and Development
Roger J. Grant

President and CEO
Perry B. Sells

Vice President, Sales and Marketing
John M. Shea

Vice President, Operations and CFO
Andrew J. Centauro

Production Manager
Helen A. Marcella

Production Coordinators
Karen L. O'Brien
Marion E. Schofield

Technical Support
Wayne D. Anderson
Thomas J. Dion
Michael H. Donelan
Gary L. Hoitt
Ivan T. Rocha
Kathryn S. Rodriguez

Book & Cover Design
Norman R. Forgit

Contractor's Pricing Guide

Framing & Rough Carpentry

1996

Published by the R.S. Means Company, Inc.

R.S. MEANS COMPANY, INC.

CONSTRUCTION
PUBLISHERS & CONSULTANTS
100 Construction Plaza, P.O. Box 800
Kingston, MA 02364-0800
(617) 585-7880

Southam
Construction
Information
Network

Foreword

Since 1942, R.S. Means Company, Inc. has been providing current and comprehensive construction cost data to construction industry professionals such as architects, engineers, owners, general contractors, and facilities managers. The *Contractor's Pricing Guide* series of publications applies that valuable experience to the building trades.

This book is designed and written to provide carpenters with professional estimating information for construction of new projects associated with the trade.

Illustrations show how basic building components are grouped together to create a complete construction "system". Various "systems" can then be used together to create specific building projects. Information listed includes material costs, man-hours required to install the system, related installation costs, and total costs per unit for each system.

Material costs are determined through information provided by manufacturers, dealers, distributors, and contractors throughout the region. These cost figures include a standard 10% mark-up for profit.

Installation costs include labor and equipment, plus a mark-up for overhead and profit.

Labor costs are based on the average of open shop wage rates across the U.S. Rates are determined from formal contracts or prevailing wages for construction trades for the current year. If wage rates in your specific area vary from those used in this book, or if rate increases are expected within a given year, labor costs should be adjusted accordingly.

Labor costs reflect productivity based on actual working conditions. These figures include time spent during a normal workday on items other than actual installation, such as material receiving and handling, reviewing drawings and specifications, mobilization, site movement, breaks, and cleanup. Productivity data is developed over an extended period so as not to be influenced by abnormal variations, and reflects a typical average.

Equipment costs include not only rental costs, but also operating costs. Equipment prices are obtained from industry sources throughout the U.S.

In addition to the cost information provided, general construction and business information is also included.

The purpose of the book is to inform you not only about costs, but also about the various factors that affect the planning and purchasing process. Pointers and precautions are provided to help you organize the project from conception through completion of construction.

Sample estimating forms and checklists show how to keep your project on track. Handy reference aids and tables reduce the number of calculations required.

With this book, you can take the first step towards organizing and planning your project, and most importantly *controlling your costs*.

Table of Contents

Table of Contents (continued)

Section One
Cost Information

This section provides the cost information necessary to make your construction estimates accurate and efficient. This information is presented in two different formats. The first and most common format is composed of a graphic illustration of the construction system and a list of the building components that are required. Several options for each system are presented. For example, the "Floor Framing Systems" pages show the cost per square foot for 2" x 8" joist framing as well as the 2" x 10" joist framing. In addition, alternate components are shown to allow you to customize the system for your specific project.

These "systems" can also be combined to create an estimate for a more complex project. For example, if you need to replace the exterior wall system below the floor framing of a second story addition, you can consult the "Exterior Wall Framing Systems" pages to calculate those costs per square foot of wall, and then add in the cost of the floor framing. In this manner, the cost for an entire framing project, matched to your specifications for size and proposed design, can be estimated.

The second format lists separately all the building components used in the various systems. This line item format is also used to list the costs for a wide variety of construction materials which may not necessarily be used in any system. For example, the "Insulation" page lists the cost per square foot for various types of building insulation.

The "How to Use" and "Sample Calculation" pages that follow define the terms that appear in the book and will help you to use the information presented.

The cost data provided in this book is both accurate and realistic for putting together an estimate. Anyone using the information provided should, however, put it into proper perspective based upon their own experience and any price fluctuations that may occur at any time in the marketplace.

A good deal of effort has been expended to break each system down into its smallest manageable tasks. Using the systems as they exist, a contractor can confidently prepare an estimate quickly and accurately knowing that the possibility of leaving something out has been eliminated.

Once an estimate has been prepared the contractor has the luxury of analyzing the system and the costs to determine if any adjustments need to be made to increase the potential of winning the job.

How To Use Cost Information

Illustration
At the top of each cost table is an illustration, brief description, and design criteria used to develop the cost.

Unit
All products represented in a particular system are defined by the industry standard.

Quantity
This is the number of line item units required for 1 system unit.

Labor-hours
Total labor-hours for a system can be found by simply multiplying the quantity of the system required times LABOR-HOURS. The resulting figure is the total labor hours needed to complete the system. (QUANTITY OF SYSTEM x LABOR-HOURS = TOTAL SYSTEM LABOR-HOURS)

Extension
Space is available for use as a work sheet. Total costs can be derived by multiplying number of system units needed x total system costs.

System Components
The components of a system are listed separately to show the user a typical makeup of that system's price. The price for each system in the table below is calculated in a similar fashion.

Unit of Measure for Each System
In the Mat., Inst., and Total columns, each cost figure is adjusted to agree with the unit of measure for the entire system.

Materials
This column contains the MATERIAL COST of each element. These cost figures include 10% for profit.

Installation
Labor rates include both the INSTALLATION COST of the contractor and the standard contractor's O&P. On the average, the LABOR COST will be 74.0% over the BARE LABOR COST.

Total
MATERIAL COST + INSTALLATION COST = TOTAL

Alternatives
Alternate components can be used to develop a custom system.

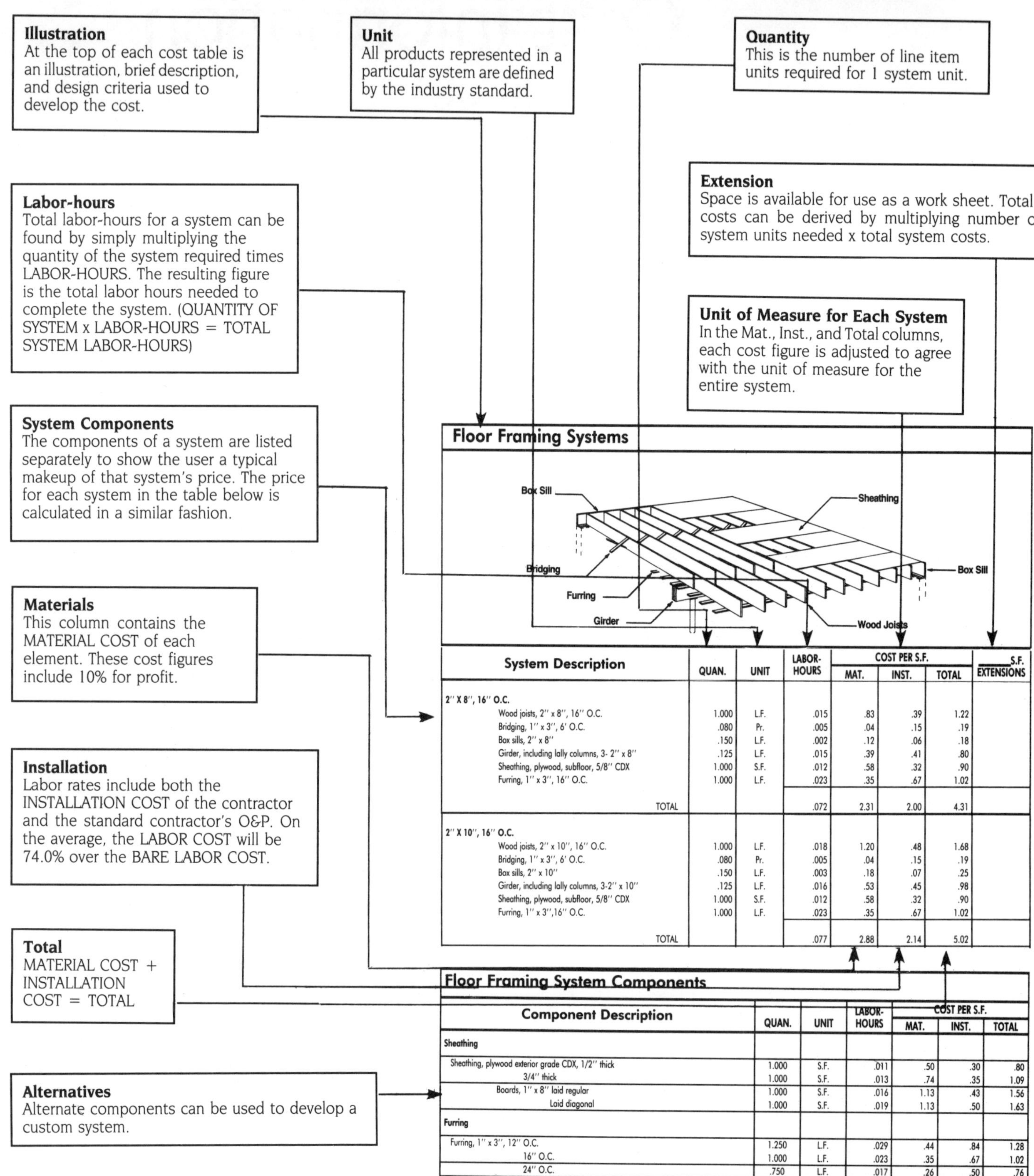

Floor Framing Systems

System Description	QUAN.	UNIT	LABOR-HOURS	COST PER S.F.			S.F. EXTENSIONS
				MAT.	INST.	TOTAL	
2″ X 8″, 16″ O.C.							
Wood joists, 2″ x 8″, 16″ O.C.	1.000	L.F.	.015	.83	.39	1.22	
Bridging, 1″ x 3″, 6′ O.C.	.080	Pr.	.005	.04	.15	.19	
Box sills, 2″ x 8″	.150	L.F.	.002	.12	.06	.18	
Girder, including lally columns, 3- 2″ x 8″	.125	L.F.	.015	.39	.41	.80	
Sheathing, plywood, subfloor, 5/8″ CDX	1.000	S.F.	.012	.58	.32	.90	
Furring, 1″ x 3″, 16″ O.C.	1.000	L.F.	.023	.35	.67	1.02	
TOTAL			.072	2.31	2.00	4.31	
2″ X 10″, 16″ O.C.							
Wood joists, 2″ x 10″, 16″ O.C.	1.000	L.F.	.018	1.20	.48	1.68	
Bridging, 1″ x 3″, 6′ O.C.	.080	Pr.	.005	.04	.15	.19	
Box sills, 2″ x 10″	.150	L.F.	.003	.18	.07	.25	
Girder, including lally columns, 3-2″ x 10″	.125	L.F.	.016	.53	.45	.98	
Sheathing, plywood, subfloor, 5/8″ CDX	1.000	S.F.	.012	.58	.32	.90	
Furring, 1″ x 3″,16″ O.C.	1.000	L.F.	.023	.35	.67	1.02	
TOTAL			.077	2.88	2.14	5.02	

Floor Framing System Components

Component Description	QUAN.	UNIT	LABOR-HOURS	COST PER S.F.		
				MAT.	INST.	TOTAL
Sheathing						
Sheathing, plywood exterior grade CDX, 1/2″ thick	1.000	S.F.	.011	.50	.30	.80
3/4″ thick	1.000	S.F.	.013	.74	.35	1.09
Boards, 1″ x 8″ laid regular	1.000	S.F.	.016	1.13	.43	1.56
Laid diagonal	1.000	S.F.	.019	1.13	.50	1.63
Furring						
Furring, 1″ x 3″, 12″ O.C.	1.250	L.F.	.029	.44	.84	1.28
16″ O.C.	1.000	L.F.	.023	.35	.67	1.02
24″ O.C.	.750	L.F.	.017	.26	.50	.76

7

Sample Calculations

A) Determine the cost of the 2nd floor framing system for the building shown in the plan at the right using 2" x 8" @ 16" O.C.

B) Compare costs for systems composed of 2" x 8" @ 16" O.C. versus 2" x 10" @ 16" O.C.

C) How do costs change for a 2" x 8" @ 16" O.C. floor framing system when 1/2" plywood sheathing is used?

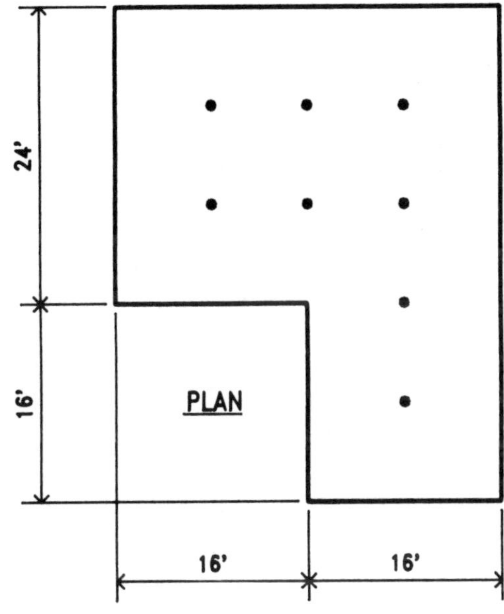

Solution:

Determine square footage of floor area to be framed. (24' x 32') + (16' x 16') = 1,024 S.F.

A) Multiply total square footage by total systems cost per S.F. 1,024 S.F. x $4.31/S.F. = $4,413.44.

B) Costs for 2" x 10" @ 16" O.C. = $5,140.48 (1,024 S.F. x 5.02/S.F.). Difference is $727.04 ($5,140.48 − $4,413.44).

C) Replace cost of 5/8" sheathing, $921.60 (1,024 S.F. x .90/S.F.) with cost of 1/2" sheathing, $819.20 (1,024 S.F. x .80/S.F.) from systems component page. New systems cost is $4,311.04 ($4,413.44 − $921.60 + $819.20).

Floor Framing Systems

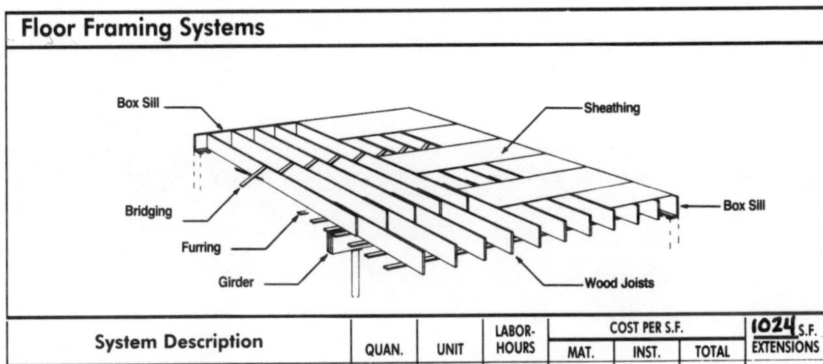

System Description	QUAN.	UNIT	LABOR-HOURS	COST PER S.F. MAT.	COST PER S.F. INST.	COST PER S.F. TOTAL	1024 S.F. EXTENSIONS
2" X 8", 16" O.C.							
Wood joists, 2" x 8", 16" O.C.	1.000	L.F.	.015	.83	.39	1.22	
Bridging, 1" x 3", 6' O.C.	.080	Pr.	.005	.04	.15	.19	
Box sills, 2" x 8"	.150	L.F.	.002	.12	.06	.18	
Girder, including lally columns, 3- 2" x 8"	.125	L.F.	.015	.39	.41	.80	
Sheathing, plywood, subfloor, 5/8" CDX	1.000	S.F.	.012	.58	.32	.90	
Furring, 1" x 3", 16" O.C.	1.000	L.F.	.023	.35	.67	1.02	922
TOTAL			.072	2.31	2.00	4.31	4413
2" X 10", 16" O.C.							
Wood joists, 2" x 10", 16" O.C.	1.000	L.F.	.018	1.20	.48	1.68	
Bridging, 1" x 3", 6' O.C.	.080	Pr.	.005	.04	.15	.19	
Box sills, 2" x 10"	.150	L.F.	.003	.18	.07	.25	
Girder, including lally columns, 3-2" x 10"	.125	L.F.	.016	.53	.45	.98	
Sheathing, plywood, subfloor, 5/8" CDX	1.000	S.F.	.012	.58	.32	.90	
Furring, 1" x 3",16" O.C.	1.000	L.F.	.023	.35	.67	1.02	
TOTAL			.077	2.88	2.14	5.02	5140

Floor Framing System Components

Component Description	QUAN.	UNIT	LABOR-HOURS	COST PER S.F. MAT.	COST PER S.F. INST.	COST PER S.F. TOTAL
Sheathing						
Sheathing, plywood exterior grade CDX, 1/2" thick	1.000	S.F.	.011	.50	.30	.80
3/4" thick	1.000	S.F.	.013	.74	.35	1.09
Boards, 1" x 8" laid regular	1.000	S.F.	.016	1.13	.43	1.56
Laid diagonal	1.000	S.F.	.019	1.13	.50	1.63
Furring						
Furring, 1" x 3", 12" O.C.	1.250	L.F.	.029	.44	.84	1.28
16" O.C.	1.000	L.F.	.023	.35	.67	1.02
24" O.C.	.750	L.F.	.017	.26	.50	.76

Framing Systems

Floor Framing Systems

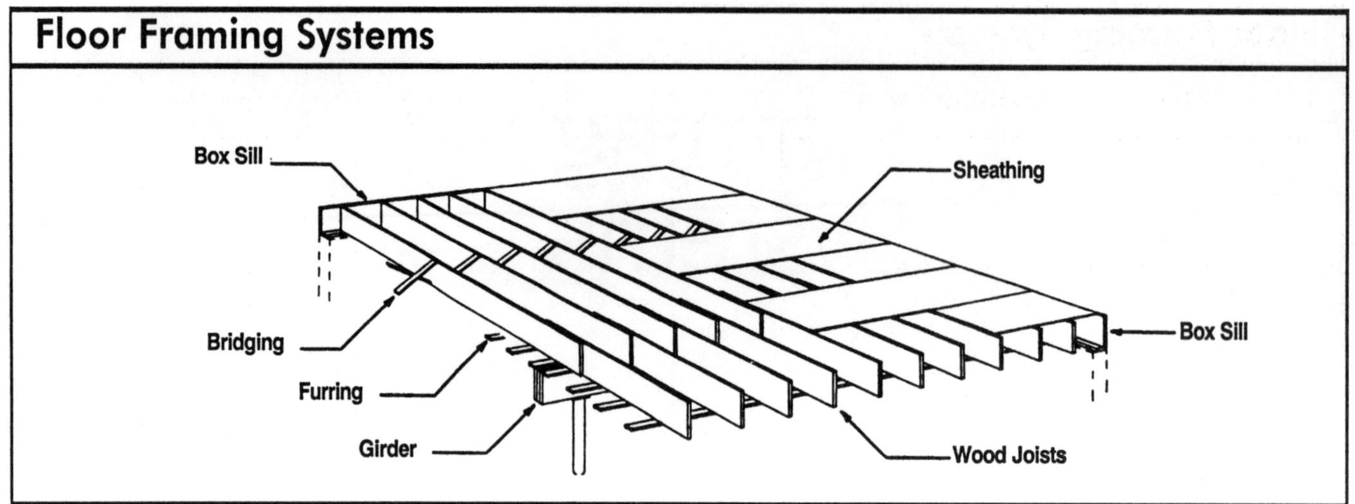

System Description	QUAN.	UNIT	LABOR-HOURS	COST PER S.F.			_____ S.F. EXTENSIONS
				MAT.	INST.	TOTAL	
2″ X 6″, 12″ O.C.							
Wood joists, 2″ x 6″, 12″ O.C.	1.250	L.F.	.016	.73	.44	1.17	
Bridging, 1″ x 3″, 6′ OC	.080	Pr.	.005	.04	.15	.19	
Box sills, 2″ x 6″	.150	L.F.	.002	.09	.05	.14	
Girder, including lally columns, 3- 2″ x 8″	.125	L.F.	.015	.39	.41	.80	
Sheathing, plywood, subfloor, 5/8″ CDX	1.000	S.F.	.012	.58	.32	.90	
Furring, 1″ x 3″, 16″ O.C.	1.000	L.F.	.023	.35	.67	1.02	
TOTAL			.073	2.18	2.04	4.22	
2″ X 6″, 16″ O.C.							
Wood joists, 2″ x 6″, 16″ O.C.	1.000	L.F.	.013	.58	.35	.93	
Bridging, 1″ x 3″, 6′ OC	.080	Pr.	.005	.04	.15	.19	
Box sills, 2″ x 6″	.150	L.F.	.002	.09	.05	.14	
Girder, including lally columns, 3- 2″ x 8″	.125	L.F.	.015	.39	.41	.80	
Sheathing, plywood, subfloor, 5/8″ CDX	1.000	S.F.	.012	.58	.32	.90	
Furring, 1″ x 3″, 16″ O.C.	1.000	L.F.	.023	.35	.67	1.02	
TOTAL			.070	2.03	1.95	3.98	
2″ X 8″, 12″ O.C.							
Wood joists, 2″ x 8″, 12″ O.C.	1.250	L.F.	.018	1.04	.49	1.53	
Bridging, 1″ x 3″, 6′ OC	.080	Pr.	.005	.04	.15	.19	
Box sills, 2″ x 8″	.150	L.F.	.002	.12	.06	.18	
Girder, including lally columns, 3- 2″ x 8″	.125	L.F.	.015	.39	.41	.80	
Sheathing, plywood, subfloor, 5/8″ CDX	1.000	S.F.	.012	.58	.32	.90	
Furring, 1″ x 3″, 16″ O.C.	1.000	L.F.	.023	.35	.67	1.02	
TOTAL			.075	2.52	2.10	4.62	
2″ X 8″, 16″ O.C.							
Wood joists, 2″ x 8″, 16″ O.C.	1.000	L.F.	.015	.83	.39	1.22	
Bridging, 1″ x 3″, 6′ O.C.	.080	Pr.	.005	.04	.15	.19	
Box sills, 2″ x 8″	.150	L.F.	.002	.12	.06	.18	
Girder, including lally columns, 3- 2″ x 8″	.125	L.F.	.015	.39	.41	.80	
Sheathing, plywood, subfloor, 5/8″ CDX	1.000	S.F.	.012	.58	.32	.90	
Furring, 1″ x 3″, 16″ O.C.	1.000	L.F.	.023	.35	.67	1.02	
TOTAL			.072	2.31	2.00	4.31	

Floor Framing Systems

System Description	QUAN.	UNIT	LABOR-HOURS	COST PER S.F. MAT.	COST PER S.F. INST.	COST PER S.F. TOTAL	S.F. EXTENSIONS
2" X 10", 12" O.C.							
Wood joists, 2" x 10", 12" O.C.	1.250	L.F.	.022	1.50	.61	2.11	
Bridging, 1" x 3", 6' OC	.080	Pr.	.005	.04	.15	.19	
Box sills, 2" x 10"	.150	L.F.	.003	.18	.07	.25	
Girder, including lally columns, 3-2" x 10"	.125	L.F.	.016	.53	.45	.98	
Sheathing, plywood, subfloor, 5/8" CDX	1.000	S.F.	.012	.58	.32	.90	
Furring, 1" x 3", 16" O.C.	1.000	L.F.	.023	.35	.67	1.02	
TOTAL			.081	3.18	2.27	5.45	
2" X 10", 16" O.C.							
Wood joists, 2" x 10", 16" O.C.	1.000	L.F.	.018	1.20	.48	1.68	
Bridging, 1" x 3", 6' O.C.	.080	Pr.	.005	.04	.15	.19	
Box sills, 2" x 10"	.150	L.F.	.003	.18	.07	.25	
Girder, including lally columns, 3-2" x 10"	.125	L.F.	.016	.53	.45	.98	
Sheathing, plywood, subfloor, 5/8" CDX	1.000	S.F.	.012	.58	.32	.90	
Furring, 1" x 3", 16" O.C.	1.000	L.F.	.023	.35	.67	1.02	
TOTAL			.077	2.88	2.14	5.02	
2" X 12", 12" O.C.							
Wood joists, 2" x 12", 12" O.C.	1.250	L.F.	.023	1.94	.62	2.56	
Bridging, 1" x 3", 6' OC	.080	Pr.	.005	.04	.15	.19	
Box sills, 2" x 12"	.150	L.F.	.003	.23	.07	.30	
Girder, including lally columns, 3-2" x 12"	.125	L.F.	.017	.67	.46	1.13	
Sheathing, plywood, subfloor, 5/8" CDX	1.000	S.F.	.012	.58	.32	.90	
Furring, 1" x 3", 16" O.C.	1.000	L.F.	.023	.35	.67	1.02	
TOTAL			.083	3.81	2.29	6.10	
2" X 12", 16" O.C.							
Wood joists, 2" x 12", 16" O.C.	1.000	L.F.	.018	1.55	.49	2.04	
Bridging, 1" x 3", 6' O.C.	.080	Pr.	.005	.04	.15	.19	
Box sills, 2" x 12"	.150	L.F.	.003	.23	.07	.30	
Girder, including lally columns, 3-2" x 12"	.125	L.F.	.017	.67	.46	1.13	
Sheathing, plywood, subfloor, 5/8" CDX	1.000	S.F.	.012	.58	.32	.90	
Furring, 1" x 3", 16" O.C.	1.000	L.F.	.023	.35	.67	1.02	
TOTAL			.078	3.42	2.16	5.58	

Floor costs on this page are given on a cost per square foot basis.

Floor Framing System Components

Component Description	QUAN.	UNIT	LABOR-HOURS	COST PER S.F. MAT.	COST PER S.F. INST.	COST PER S.F. TOTAL
Sheathing						
Sheathing, plywood exterior grade CDX, 1/2" thick	1.000	S.F.	.011	.50	.30	.80
3/4" thick	1.000	S.F.	.013	.74	.35	1.09
Boards, 1" x 8" laid regular	1.000	S.F.	.016	1.13	.43	1.56
Laid diagonal	1.000	S.F.	.019	1.13	.50	1.63
Furring						
Furring, 1" x 3", 12" O.C.	1.250	L.F.	.029	.44	.84	1.28
16" O.C.	1.000	L.F.	.023	.35	.67	1.02
24" O.C.	.750	L.F.	.017	.26	.50	.76

Floor Framing Systems

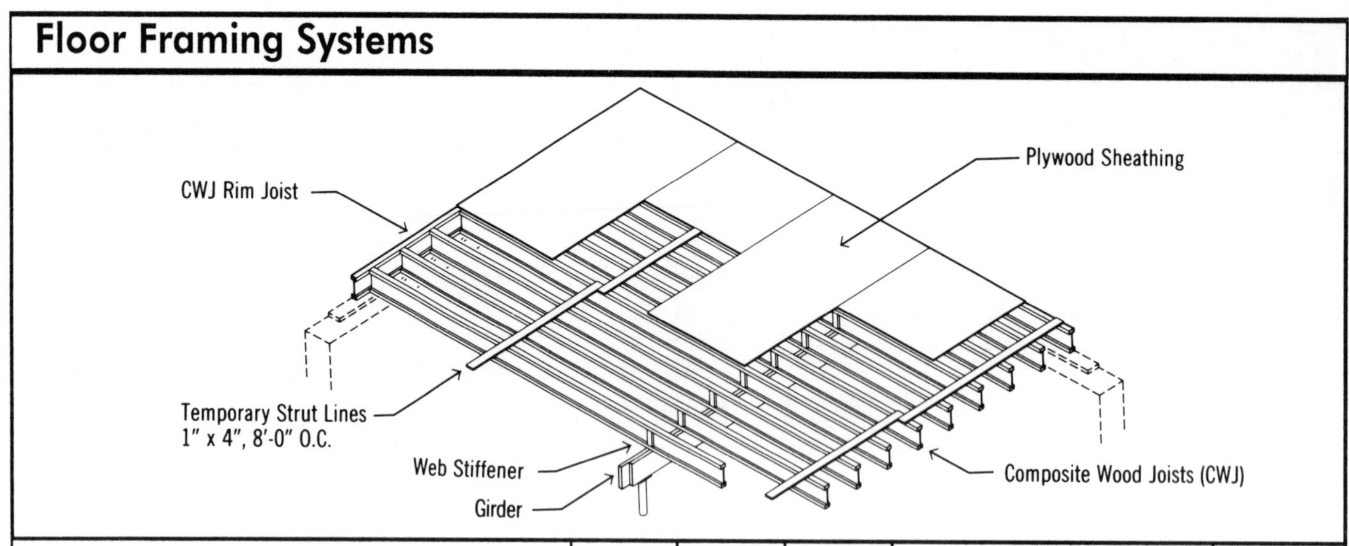

System Description	QUAN.	UNIT	LABOR-HOURS	COST PER S.F.			S.F. EXTENSIONS
				MAT.	INST.	TOTAL	
9-1/2" COMPOSITE WOOD JOISTS, 16" O.C.							
CWJ, 9-1/2", 16" O.C., 15' span	1.000	L.F.	.018	1.65	.48	2.13	
Temp. strut line, 1" x 4", 8' O.C.	.160	L.F.	.003	.07	.08	.15	
CWJ rim joist, 9-1/2"	.150	L.F.	.003	.25	.07	.32	
Girder, including lally columns, 3- 2" x 8"	.125	L.F.	.015	.39	.41	.80	
Sheathing, plywood, subfloor, 5/8" CDX	1.000	S.F.	.012	.58	.32	.90	
TOTAL			.051	2.94	1.36	4.30	
9-1/2" COMPOSITE WOOD JOISTS, 24" O.C.							
CWJ, 9-1/2", 24" O.C., 13' span	.750	L.F.	.013	1.24	.36	1.60	
Temp. strut line, 1" x 4", 8' OC	.160	L.F.	.003	.07	.08	.15	
CWJ rim joist, 9-1/2"	.150	L.F.	.003	.25	.07	.32	
Girder, including lally columns, 3- 2" x 10"	.125	L.F.	.015	.39	.41	.80	
Sheathing, plywood, subfloor, 5/8" CDX	1.000	S.F.	.012	.58	.32	.90	
TOTAL			.046	2.53	1.24	3.77	
11-1/2" COMPOSITE WOOD JOISTS, 16" O.C.							
CWJ, 11-1/2", 16" O.C., 18' span	1.000	L.F.	.018	2.03	.49	2.52	
Temp. strut line, 1" x 4", 8' O.C.	.160	L.F.	.003	.07	.08	.15	
CWJ rim joist, 11-1/2"	.150	L.F.	.003	.30	.07	.37	
Girder, including lally columns, 3-2" x 10"	.125	L.F.	.016	.53	.45	.98	
Sheathing, plywood, subfloor, 5/8" CDX	1.000	S.F.	.012	.58	.32	.90	
TOTAL			.052	3.51	1.41	4.92	
11-1/2" COMPOSITE WOOD JOISTS, 24" O.C.							
CWJ, 11-1/2", 24" O.C., 16' span	.750	L.F.	.014	1.52	.37	1.89	
Temp. strut line, 1" x 4", 8' OC	.160	L.F.	.003	.07	.08	.15	
CWJ rim joist, 11-1/2"	.150	L.F.	.003	.30	.07	.37	
Girder, including lally columns, 3- 2" x 10"	.125	L.F.	.016	.53	.45	.98	
Sheathing, plywood, subfloor, 5/8" CDX	1.000	S.F.	.012	.58	.32	.90	
TOTAL			.048	3.00	1.29	4.29	

Floor Framing Systems

System Description	QUAN.	UNIT	LABOR-HOURS	MAT.	INST.	TOTAL	S.F. EXTENSIONS
14″ COMPOSITE WOOD JOISTS, 16″ O.C.							
CWJ, 14″, 16″ O.C., 22′ span	1.000	L.F.	.020	2.65	.53	3.18	
Temp. strut line, 1″ x 4″, 8′ O.C.	.160	L.F.	.003	.07	.08	.15	
CWJ rim joist, 14″	.150	L.F.	.003	.40	.08	.48	
Girder, including lally columns, 3-2″ x 12″	.600	L.F.	.017	.67	.46	1.13	
Sheathing, plywood, subfloor, 5/8″ CDX	1.000	S.F.	.012	.58	.32	.90	
TOTAL			.055	4.37	1.47	5.84	
14″ COMPOSITE WOOD JOISTS, 24″ O.C.							
CWJ, 14″, 24″ O.C., 19′ span	.750	L.F.	.015	1.99	.40	2.39	
Temp. strut line, 1″ x 4″, 8′ OC	.160	L.F.	.003	.07	.08	.15	
CWJ rim joist, 14″	.150	L.F.	.003	.40	.08	.48	
Girder, including lally columns, 3- 2″ x 12″	.060	L.F.	.017	.67	.46	1.13	
Sheathing, plywood, subfloor, 5/8″ CDX	1.000	S.F.	.012	.58	.32	.90	
TOTAL			.050	3.71	1.34	5.05	
16″ COMPOSITE WOOD JOISTS, 16″ O.C.							
CWJ, 16″, 16″ O.C., 24′ span	1.000	L.F.	.018	1.65	.48	2.13	
Temp. strut line, 1″ x 4″, 8′ OC	.160	L.F.	.003	.07	.08	.15	
CWJ rim joist, 16″	.150	L.F.	.003	.43	.08	.51	
Girder, including lally columns, 3- 2″ x 12″	.060	L.F.	.017	.67	.46	1.13	
Sheathing, plywood, subfloor, 5/8″ CDX	1.000	S.F.	.012	.58	.32	.90	
TOTAL			.053	3.40	1.42	4.82	
16″ COMPOSITE WOOD JOISTS, 24″ O.C.							
CWJ, 16″, 24″ O.C., 22′ span	.750	L.F.	.013	1.24	.36	1.60	
Temp. strut line, 1″ x 4″, 8′ OC	.160	L.F.	.003	.07	.08	.15	
CWJ rim joist, 16″	.150	L.F.	.003	.43	.08	.51	
Girder, including lally columns, 3- 2″ x 12″	.060	L.F.	.017	.67	.46	1.13	
Sheathing, plywood, subfloor, 5/8″ CDX	1.000	S.F.	.012	.58	.32	.90	
TOTAL			.048	2.99	1.30	4.29	

Floor costs on this page are given on a cost per square foot basis.

Floor Framing System Components

Component Description	QUAN.	UNIT	LABOR-HOURS	MAT.	INST.	TOTAL
Sheathing						
Sheathing, plywood exterior grade CDX, 1/2″ thick	1.000	S.F.	.011	.50	.30	.80
5/8″ thick	1.000	S.F.	.012	.58	.32	.90
3/4″ thick	1.000	S.F.	.013	.74	.35	1.09
Boards, 1″ x 8″ laid regular	1.000	S.F.	.016	1.13	.43	1.56
Laid diagonal	1.000	S.F.	.019	1.13	.50	1.63
1″ x 10″ laid regular	1.000	S.F.	.015	1.25	.39	1.64
Laid diagonal	1.000	S.F.	.018	1.25	.48	1.73
Furring, 1″ x 3″, 12″ O.C.	1.250	L.F.	.029	.44	.84	1.28
16″ O.C.	1.000	L.F.	.023	.35	.67	1.02
24″ O.C.	.750	L.F.	.017	.26	.50	.76

Floor Framing Systems

System Description	QUAN.	UNIT	LABOR-HOURS	COST PER S.F.			S.F. EXTENSIONS
				MAT.	INST.	TOTAL	
12″ OPEN WEB JOISTS, 16″ O.C.							
OWJ 12″, 16″ O.C., 21′ span	1.000	L.F.	.018	2.35	.49	2.84	
Continuous ribbing, 2″ x 4″	.150	L.F.	.002	.06	.05	.11	
Girder, including lally columns, 3- 2″ x 8″	.125	L.F.	.015	.39	.41	.80	
Sheathing, plywood, subfloor, 5/8″ CDX	1.000	S.F.	.012	.58	.32	.90	
Furring, 1″ x 3″, 16″ O.C.	1.000	L.F.	.023	.35	.67	1.02	
TOTAL			.070	3.73	1.94	5.67	
12″ OPEN WEB WOOD JOISTS, 24″ O.C.							
OWJ, 12″, 24″ O.C., 17′ span	.750	L.F.	.014	1.76	.37	2.13	
Continuous ribbing, 2″ x 4″	.150	L.F.	.002	.06	.05	.11	
Girder, including lally columns, 3-2″ x 10″	.125	L.F.	.015	.39	.41	.80	
Sheathing, plywood, subfloor, 5/8″ CDX	1.000	S.F.	.012	.58	.32	.90	
Furring, 1″ x 3″, 16″ O.C.	1.000	L.F.	.023	.35	.67	1.02	
TOTAL			.066	3.14	1.82	4.96	
14″ OPEN WEB WOOD JOISTS, 16″ O.C.							
OWJ 14″, 16″ O.C., 22′ span	1.000	L.F.	.020	2.58	.53	3.11	
Continuous ribbing, 2″ x 4″	.150	L.F.	.002	.06	.05	.11	
Girder, including lally columns, 3-2″ x 10″	.125	L.F.	.016	.53	.45	.98	
Sheathing, plywood, subfloor, 5/8″ CDX	1.000	S.F.	.012	.58	.32	.90	
Furring, 1″ x 3″,16″ O.C.	1.000	L.F.	.023	.35	.67	1.02	
TOTAL			.073	4.10	2.02	6.12	
14″ OPEN WEB WOOD JOISTS, 24″ O.C.							
OWJ 14″, 24″ O.C., 18′ span	.750	L.F.	.015	1.93	.40	2.33	
Continuous ribbing, 2″ x 4″	.150	L.F.	.002	.06	.05	.11	
Girder, including lally columns, 3-2″ x 10″	.125	L.F.	.016	.53	.45	.98	
Sheathing, plywood, subfloor, 5/8″ CDX	1.000	S.F.	.012	.58	.32	.90	
Furring, 1″ x 3″,16″ O.C.	1.000	L.F.	.023	.35	.67	1.02	
TOTAL			.068	3.45	1.89	5.34	

Floor Framing Systems

System Description	QUAN.	UNIT	LABOR-HOURS	MAT.	INST.	TOTAL	S.F. EXTENSIONS
16″ OPEN WEB WOOD JOISTS, 16″ O.C.							
OWJ 16″, 16″ O.C., 24′ span	1.000	L.F.	.021	2.70	.56	3.26	
Continuous ribbing, 2″ x 4″	.150	L.F.	.002	.06	.05	.11	
Girder, including lally columns, 3-2″ x 12″	.125	L.F.	.017	.67	.46	1.13	
Sheathing, plywood, subfloor, 5/8″ CDX	1.000	S.F.	.012	.58	.32	.90	
Furring, 1″ x 3″, 16″ O.C.	1.000	L.F.	.023	.35	.67	1.02	
TOTAL			.075	4.36	2.06	6.42	
16″ OPEN WEB WOOD JOISTS, 24″ O.C.							
OWJ 16″, 24″ O.C., 20′ span	.750	L.F.	.015	2.03	.41	2.44	
Continuous ribbing, 2″ x 4″	.150	L.F.	.002	.06	.05	.11	
Girder, including lally columns, 3-2″ x 12″	.125	L.F.	.017	.67	.46	1.13	
Sheathing, plywood, subfloor, 5/8″ CDX	1.000	S.F.	.012	.58	.32	.90	
Furring, 1″ x 3″, 16″ O.C.	1.000	L.F.	.023	.35	.67	1.02	
TOTAL			.069	3.69	1.91	5.60	
18″ OPEN WEB WOOD JOISTS, 16″ O.C.							
OWJ 18″, 16″ O.C., 24′ span	1.000	L.F.	.022	2.80	.59	3.39	
Continuous ribbing, 2″ x 4″	.150	L.F.	.002	.06	.05	.11	
Girder, including lally columns, 3-2″ x 12″	.125	L.F.	.017	.67	.46	1.13	
Sheathing, plywood, subfloor, 5/8″ CDX	1.000	S.F.	.012	.58	.32	.90	
Furring, 1″ x 3″, 16″ O.C.	1.000	L.F.	.023	.35	.67	1.02	
TOTAL			.076	4.46	2.09	6.55	
18″ OPEN WEB WOOD JOISTS, 24″ O.C.							
OWJ 18″, 24″ O.C., 20′ span	.750	L.F.	.016	2.10	.44	2.54	
Continuous ribbing, 2″ x 4″	.150	L.F.	.002	.06	.05	.11	
Girder, including lally columns, 3-2″ x 12″	.125	L.F.	.017	.67	.46	1.13	
Sheathing, plywood, subfloor, 5/8″ CDX	1.000	S.F.	.012	.58	.32	.90	
Furring, 1″ x 3″, 16″ O.C.	1.000	L.F.	.023	.35	.67	1.02	
TOTAL			.070	3.76	1.94	5.70	

Floor costs on this page are given on a cost per square foot basis.

Floor Framing System Components

Component Description	QUAN.	UNIT	LABOR-HOURS	MAT.	INST.	TOTAL
Sheathing						
Sheathing, plywood exterior grade CDX, 1/2″ thick	1.000	S.F.	.011	.50	.30	.80
3/4″ thick	1.000	S.F.	.013	.74	.35	1.09
Furring						
Furring, 1″ x 3″, 12″ O.C.	1.250	L.F.	.029	.44	.84	1.28
24″ O.C.	.750	L.F.	.017	.26	.50	.76

Exterior Wall Framing Systems

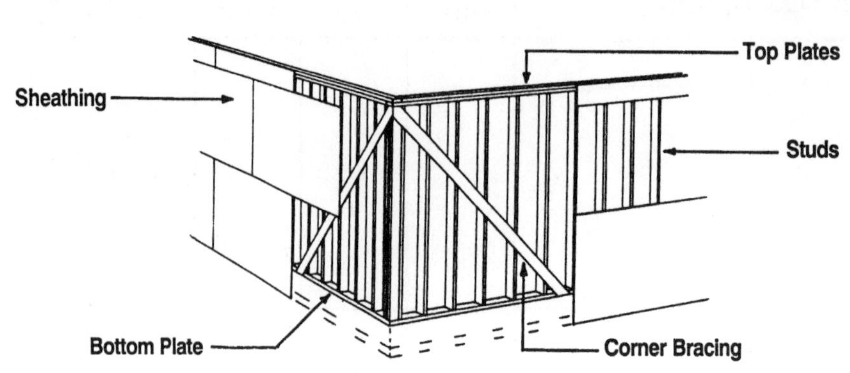

System Description	QUAN.	UNIT	LABOR-HOURS	COST PER S.F.			S.F. EXTENSIONS
				MAT.	INST.	TOTAL	
2″ x 4″, 12″ O.C.							
2″ x 4″ studs, 12″ O.C.	1.250	L.F.	.018	.50	.49	.99	
Plates, 2″ x 4″, double top, single bottom	.375	L.F.	.005	.15	.15	.30	
Corner bracing, let-in, 1″ x 6″	.063	L.F.	.003	.03	.10	.13	
Sheathing, 1/2″ plywood, CDX	1.000	S.F.	.011	.50	.30	.80	
TOTAL			.037	1.18	1.04	2.22	
2″ X 4″, 16″ O.C.							
2″ x 4″ studs, 16″ O.C.	1.000	L.F.	.015	.40	.39	.79	
Plates, 2″ x 4″, double top, single bottom	.375	L.F.	.005	.15	.15	.30	
Corner bracing, let-in, 1″ x 6″	.063	L.F.	.003	.03	.10	.13	
Sheathing, 1/2″ plywood, CDX	1.000	S.F.	.011	.50	.30	.80	
TOTAL			.034	1.08	.94	2.02	
2″ X 4″, 24″ O.C.							
2″ x 4″ studs, 24″ O.C.	.750	L.F.	.011	.30	.30	.60	
Plates, 2″ x 4″, double top, single bottom	.375	L.F.	.005	.15	.15	.30	
Corner bracing, let-in, 1″ x 6″	.063	L.F.	.002	.03	.06	.09	
Sheathing, 1/2″ plywood, CDX	1.000	S.F.	.011	.50	.30	.80	
TOTAL			.029	.98	.81	1.79	
2″ x 6″, 12″ O.C.							
2″ x 6″ studs, 12″ O.C.	1.250	L.F.	.020	.73	.54	1.27	
Plates, 2″ x 6″, double top, single bottom	.375	L.F.	.006	.22	.16	.38	
Corner bracing, let-in, 1″ x 6″	.063	L.F.	.003	.03	.10	.13	
Sheathing, 1/2″ plywood, CDX	1.000	S.F.	.011	.50	.30	.80	
TOTAL			.040	1.48	1.10	2.58	
2″ X 6″, 16″ O.C.							
2″ x 6″ studs, 16″ O.C.	1.000	L.F.	.016	.58	.43	1.01	
Plates, 2″ x 6″, double top, single bottom	.375	L.F.	.006	.22	.16	.38	
Corner bracing, let-in, 1″ x 6″	.063	L.F.	.003	.03	.10	.13	
Sheathing, 1/2″ plywood, CDX	1.000	S.F.	.011	.50	.30	.80	
TOTAL			.036	1.33	.99	2.32	

Exterior Wall Framing Systems

System Description	QUAN.	UNIT	LABOR-HOURS	COST PER S.F.			S.F. EXTENSIONS
				MAT.	INST.	TOTAL	
2″ X 6″, 24″ O.C.							
2″ x 6″ studs, 24″ O.C.	.750	L.F.	.012	.44	.33	.77	
Plates, 2″ x 6″, double top, single bottom	.375	L.F.	.006	.22	.16	.38	
Corner bracing, let-in, 1″ x 6″	.063	L.F.	.002	.03	.06	.09	
Sheathing, 1/2″ plywood, CDX	1.000	S.F.	.011	.50	.30	.80	
TOTAL			.031	1.19	.85	2.04	
2″ x 8″, 16″ O.C.							
2″ x 8″ studs, 16″ O.C.	1.000	L.F.	.020	1.32	.54	1.86	
Plates, 2″ x 8″, double top, single bottom	.375	L.F.	.008	.50	.20	.70	
Corner bracing, let-in, 1″ x 6″	.063	L.F.	.002	.03	.06	.09	
Sheathing, 1/2″ plywood, CDX	1.000	S.F.	.011	.50	.30	.80	
TOTAL			.041	2.35	1.10	3.45	
2″ x 8″, 24″ O.C.							
2″ x 8″ studs, 24″ O.C.	.750	L.F.	.015	.99	.40	1.39	
Plates, 2″ x 8″, double top, single bottom	.375	L.F.	.008	.50	.20	.70	
Corner bracing, let-in, 1″ x 6″	.063	L.F.	.002	.03	.06	.09	
Sheathing, 1/2″ plywood, CDX	1.000	S.F.	.011	.50	.30	.80	
TOTAL			.036	2.02	.96	2.98	

The wall costs on this page are given in cost per square foot of wall.
For window and door openings see below.

Exterior Wall Framing Components

Component Description	QUAN.	UNIT	LABOR-HOURS	COST PER S.F.		
				MAT.	INST.	TOTAL
Sheathing						
Sheathing, plywood CDX, 3/8″ thick	1.000	S.F.	.010	.40	.28	.68
5/8″ thick	1.000	S.F.	.012	.58	.34	.92
3/4″ thick	1.000	S.F.	.013	.74	.36	1.10
Wood fiber, regular, no vapor barrier, 1/2″ thick	1.000	S.F.	.013	.41	.36	.77
5/8″ thick	1.000	S.F.	.013	.54	.36	.90
Asphalt impregnated 25/32″ thick	1.000	S.F.	.013	.36	.36	.72
1/2″ thick	1.000	S.F.	.013	.30	.36	.66

Window & Door Openings

Component Description	QUAN.	UNIT	LABOR-HOURS	COST EACH		
				MAT.	INST.	TOTAL
The following costs are to be added to the total costs of the wall for each opening. Do not subtract the area of the openings.						
Headers						
Headers, 2″ x 6″ double, 2′ long	4.000	L.F.	.178	2.32	4.80	7.12
4′ long	8.000	L.F.	.356	4.64	9.60	14.24
2″ x 8″ double, 4′ long	8.000	L.F.	.376	6.65	10.20	16.85
6′ long	12.000	L.F.	.565	9.95	15.20	25.15
2″ x 10″ double, 4′ long	8.000	L.F.	.400	9.60	10.75	20.35
6′ long	12.000	L.F.	.600	14.40	16.05	30.45

Gable End Roof Framing Systems

System Description	QUAN.	UNIT	LABOR-HOURS	COST PER S.F.			S.F. EXTENSIONS
				MAT.	INST.	TOTAL	
2'' X 6'' RAFTERS, 16'' O.C., 4/12 PITCH							
Rafters, 2'' x 6'', 16'' O.C., 4/12 pitch	1.170	L.F.	.019	.68	.50	1.18	
Ceiling joists, 2'' x 4'', 16'' O.C.	1.000	L.F.	.013	.40	.35	.75	
Ridge board, 2'' x 6''	.050	L.F.	.002	.03	.04	.07	
Fascia board, 2'' x 6''	.100	L.F.	.005	.06	.15	.21	
Rafter tie, 1'' x 4'', 4' O.C.	.060	L.F.	.001	.03	.03	.06	
Soffit nailer (outrigger), 2'' x 4'', 24'' O.C.	.170	L.F.	.004	.07	.12	.19	
Sheathing, exterior, plywood, CDX, 1/2'' thick	1.170	S.F.	.013	.59	.35	.94	
Furring strips, 1'' x 3'', 16'' O.C.	1.000	L.F.	.023	.35	.67	1.02	
TOTAL			.080	2.21	2.21	4.42	
2'' X 8'' RAFTERS, 16'' O.C., 4/12 PITCH							
Rafters, 2'' x 8'', 16'' O.C., 4/12 pitch	1.170	L.F.	.020	.97	.52	1.49	
Ceiling joists, 2'' x 6'', 16'' O.C.	1.000	L.F.	.013	.58	.35	.93	
Ridge board, 2'' x 8''	.050	L.F.	.002	.04	.05	.09	
Fascia board, 2'' x 8''	.100	L.F.	.007	.08	.19	.27	
Rafter tie, 1'' x 4'', 4' O.C.	.060	L.F.	.001	.03	.03	.06	
Soffit nailer (outrigger), 2'' x 4'', 24'' O.C.	.170	L.F.	.004	.07	.12	.19	
Sheathing, exterior, plywood, CDX, 1/2'' thick	1.170	S.F.	.013	.59	.35	.94	
Furring strips, 1'' x 3'', 16'' O.C.	1.000	L.F.	.023	.35	.67	1.02	
TOTAL			.083	2.71	2.28	4.99	
2'' X 8'' RAFTERS, 24'' O.C., 4/12 PITCH							
Rafters, 2'' x 8'', 24'' O.C., 4/12 pitch	.940	L.F.	.016	.78	.42	1.20	
Ceiling joists, 2'' x 6'', 24'' O.C.	.750	L.F.	.010	.44	.27	.71	
Ridge board, 2'' x 8''	.050	L.F.	.002	.03	.04	.07	
Fascia board, 2'' x 8''	.100	L.F.	.005	.06	.15	.21	
Rafter tie, 1'' x 4'', 4' OC	.060	L.F.	.001	.03	.03	.06	
Soffit nailer (outrigger), 2'' x 4'', 24'' O.C.	.170	L.F.	.004	.07	.12	.19	
Sheathing, exterior, plywood, CDX, 1/2'' thick	1.170	S.F.	.013	.59	.35	.94	
Furring strips, 1'' x 3'', 16'' O.C.	1.000	L.F.	.023	.35	.67	1.02	
TOTAL			.074	2.35	2.05	4.40	

Gable End Roof Framing Systems

System Description	QUAN.	UNIT	LABOR-HOURS	COST PER S.F. MAT.	COST PER S.F. INST.	COST PER S.F. TOTAL	S.F. EXTENSIONS
2″ X 10″ RAFTERS, 16″ O.C., 4/12 PITCH							
Rafters, 2″ x 10″, 16″ O.C., 4/12 pitch	1.170	L.F.	.030	1.40	.80	2.20	
Ceiling joists, 2″ x 8″, 16″ O.C.	1.000	L.F.	.015	.83	.39	1.22	
Ridge board, 2″ x 10″	.050	L.F.	.002	.03	.04	.07	
Fascia board, 2″ x 10″	.100	L.F.	.005	.06	.15	.21	
Rafter tie, 1″ x 4″, 4′ OC	.060	L.F.	.001	.03	.03	.06	
Soffit nailer (outrigger), 2″ x 4″, 24″ O.C.	.170	L.F.	.004	.07	.12	.19	
Sheathing, exterior, plywood, CDX, 1/2″ thick	1.170	S.F.	.013	.59	.35	.94	
Furring strips, 1″ x 3″, 16″ O.C.	1.000	L.F.	.023	.35	.67	1.02	
TOTAL			.093	3.36	2.55	5.91	
2″ X 10″ RAFTERS, 24″ O.C., 4/12 PITCH							
Rafters, 2″ x 10″, 24″ O.C., 4/12 pitch	.940	L.F.	.024	1.13	.64	1.77	
Ceiling joists, 2″ x 8″, 24″ O.C.	.750	L.F.	.011	.62	.30	.92	
Ridge board, 2″ x 10″	.050	L.F.	.002	.03	.04	.07	
Fascia board, 2″ x 10″	.100	L.F.	.005	.06	.15	.21	
Rafter tie, 1″ x 4″, 4′ OC	.060	L.F.	.001	.03	.03	.06	
Soffit nailer (outrigger), 2″ x 4″, 24″ O.C.	.170	L.F.	.004	.07	.12	.19	
Sheathing, exterior, plywood, CDX, 1/2″ thick	1.170	S.F.	.013	.59	.35	.94	
Furring strips, 1″ x 3″, 16″ O.C.	1.000	L.F.	.023	.35	.67	1.02	
TOTAL			.083	2.88	2.30	5.18	

The cost of this system is based on the square foot of plan area.
All quantities have been adjusted accordingly.

Gable End Roof Framing Components

Component Description	QUAN.	UNIT	LABOR-HOURS	COST PER S.F. MAT.	COST PER S.F. INST.	COST PER S.F. TOTAL
Rafters						
Rafters, #2 or better, 16″ O.C., 2″ x 6″, 4/12 pitch	1.170	L.F.	.019	.68	.50	1.18
8/12 pitch	1.330	L.F.	.027	.77	.72	1.49
2″ x 8″, 4/12 pitch	1.170	L.F.	.020	.97	.52	1.49
8/12 pitch	1.330	L.F.	.028	1.10	.77	1.87
24″ O.C., 2″ x 6″, 4/12 pitch	.940	L.F.	.015	.55	.41	.96
8/12 pitch	1.060	L.F.	.021	.61	.57	1.18
2″ x 8″, 4/12 pitch	.940	L.F.	.016	.78	.42	1.20
8/12 pitch	1.060	L.F.	.023	.88	.61	1.49
Ceiling joists						
Ceiling joist, #2 or better, 2″ x 4″, 16″ O.C.	1.000	L.F.	.013	.40	.35	.75
24″ O.C.	.750	L.F.	.010	.30	.27	.57
2″ x 10″, 16″ O.C.	1.000	L.F.	.018	1.20	.48	1.68
24″ O.C.	.750	L.F.	.013	.90	.37	1.27
Ridge board						
Ridge board, #2 or better, 1″ x 6″	.050	L.F.	.001	.03	.03	.06
1″ x 8″	.050	L.F.	.001	.04	.04	.08
1″ x 10″	.050	L.F.	.002	.05	.04	.09
2″ x 6″	.050	L.F.	.002	.03	.04	.07
Fascia board						
Fascia board, #2 or better, 1″ x 6″	.100	L.F.	.004	.05	.10	.15
1″ x 8″	.100	L.F.	.005	.05	.13	.18

Truss Roof Framing Systems

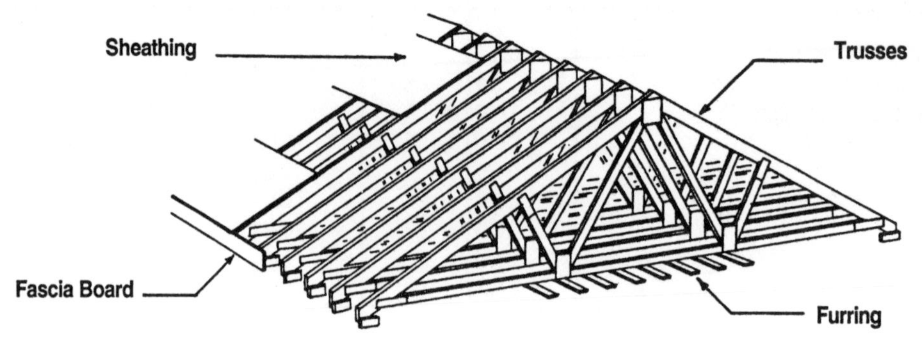

System Description	QUAN.	UNIT	LABOR-HOURS	COST PER S.F.			S.F. EXTENSIONS
				MAT.	INST.	TOTAL	
TRUSS, 16″ O.C., 4/12 PITCH, 1′ OVERHANG, 26′ SPAN							
Truss, 40# loading, 16″ O.C., 4/12 pitch, 26′ span	.030	Ea.	.021	2.04	.81	2.85	
Fascia board, 2″ x 6″	.100	L.F.	.005	.06	.15	.21	
Sheathing, exterior, plywood, CDX, 1/2″ thick	1.170	S.F.	.013	.59	.35	.94	
Furring, 1″ x 3″, 16″ O.C.	1.000	L.F.	.023	.35	.67	1.02	
TOTAL			.062	3.04	1.98	5.02	
TRUSS, 16″ O.C., 8/12 PITCH, 1′ OVERHANG, 26′ SPAN							
Truss, 40# loading, 16″ O.C., 8/12 pitch, 26′ span	.030	Ea.	.023	2.42	.88	3.30	
Fascia board, 2″ x 6″	.100	L.F.	.005	.06	.15	.21	
Sheathing, exterior, plywood, CDX, 1/2″ thick	1.330	S.F.	.015	.67	.40	1.07	
Furring, 1″ x 3″, 16″ O.C.	1.000	L.F.	.023	.35	.67	1.02	
TOTAL			.066	3.50	2.10	5.60	
TRUSS, 24″ O.C., 4/12 PITCH, 1′ OVERHANG, 26′ SPAN							
Truss, 40# loading, 24″ O.C., 4/12 pitch, 26′ span	.020	Ea.	.014	1.36	.54	1.90	
Fascia board, 2″ x 6″	.100	L.F.	.005	.06	.15	.21	
Sheathing, exterior, plywood, CDX, 1/2″ thick	1.170	S.F.	.013	.59	.35	.94	
Furring, 1″ x 3″, 16″ O.C.	1.000	L.F.	.023	.35	.67	1.02	
TOTAL			.055	2.36	1.71	4.07	
TRUSS, 24″ O.C., 8/12 PITCH, 1′ OVERHANG, 26′ SPAN							
Truss, 40# loading, 24″ O.C., 8/12 pitch, 26′ span	.020	Ea.	.015	1.61	.59	2.20	
Fascia board, 2″ x 6″	.100	L.F.	.005	.06	.15	.21	
Sheathing, exterior, plywood, CDX, 1/2″ thick	1.330	S.F.	.015	.67	.40	1.07	
Furring, 1″ x 3″, 16″ O.C.	1.000	L.F.	.023	.35	.67	1.02	
TOTAL			.058	2.69	1.81	4.50	
TRUSS, 16″ O.C., 4/12 PITCH, 1′ OVERHANG, 32′ SPAN							
Truss, 40# loading, 16″ O.C., 4/12 pitch, 32′ span	.024	Ea.	.019	2.03	.73	2.76	
Fascia board, 2″ x 6″	.100	L.F.	.005	.06	.15	.21	
Sheathing, exterior, plywood, CDX, 1/2″ thick	1.170	S.F.	.015	.67	.40	1.07	
Furring, 1″ x 3″, 16″ O.C.	1.000	L.F.	.023	.35	.67	1.02	
TOTAL			.062	3.11	1.95	5.06	

Truss Roof Framing Systems

System Description	QUAN.	UNIT	LABOR-HOURS	COST PER S.F.			S.F. EXTENSIONS
				MAT.	INST.	TOTAL	
TRUSS, 16″ O.C., 8/12 PITCH, 1′ OVERHANG, 32′ SPAN							
Truss, 40# loading, 16″ O.C., 8/12 pitch, 32′ span	.024	Ea.	.021	2.35	.82	3.17	
Fascia board, 2″ x 6″	.100	L.F.	.005	.06	.15	.21	
Sheathing, exterior, plywood, CDX, 1/2″ thick	1.330	S.F.	.015	.67	.40	1.07	
Furring, 1″ x 3″, 16″ O.C.	1.000	L.F.	.023	.35	.67	1.02	
TOTAL			.064	3.43	2.04	5.47	
TRUSS, 24″ O.C., 4/12 PITCH, 1′ OVERHANG, 32′ SPAN							
Truss, 40# loading, 24″ O.C., 4/12 pitch, 32′ span	.016	Ea.	.013	1.35	.50	1.85	
Fascia board, 2″ x 6″	.100	L.F.	.005	.06	.15	.21	
Sheathing, exterior, plywood, CDX, 1/2″ thick	1.170	S.F.	.015	.67	.40	1.07	
Furring, 1″ x 3″, 16″ O.C.	1.000	L.F.	.023	.35	.67	1.02	
TOTAL			.056	2.43	1.72	4.15	
TRUSS, 24″ O.C., 8/12 PITCH, 1′ OVERHANG, 32′ SPAN							
Truss, 40# loading, 24″ O.C., 8/12 pitch, 32′ span	.016	Ea.	.014	1.57	.55	2.12	
Fascia board, 2″ x 6″	.100	L.F.	.005	.06	.15	.21	
Sheathing, exterior, plywood, CDX, 1/2″ thick	1.330	S.F.	.015	.67	.40	1.07	
Furring, 1″ x 3″, 16″ O.C.	1.000	L.F.	.023	.35	.67	1.02	
TOTAL			.057	2.65	1.77	4.42	

The cost of this system is based on the square foot of plan area.
A one foot overhang is included.

Truss Roof Framing System Components

Component Description	QUAN.	UNIT	LABOR-HOURS	COST PER S.F.		
				MAT.	INST.	TOTAL
Trusses						
Truss, 40# loading, including 1′ overhang, 4/12 pitch, 24′ span, 16″ O.C.	.033	Ea.	.022	2.15	.85	3.00
24″ O.C.	.022	Ea.	.015	1.43	.56	1.99
28′ span, 16″ O.C.	.027	Ea.	.020	2.05	.78	2.83
24″ O.C.	.019	Ea.	.014	1.44	.55	1.99
36′ span, 16″ O.C.	.022	Ea.	.019	2.13	.74	2.87
24″ O.C.	.015	Ea.	.013	1.46	.51	1.97
8/12 pitch, 24′ span, 16″ O.C.	.033	Ea.	.024	2.48	.92	3.40
24″ O.C.	.022	Ea.	.016	1.65	.61	2.26
28′ span, 16″ O.C.	.027	Ea.	.022	2.35	.85	3.20
24″ O.C.	.019	Ea.	.016	1.65	.60	2.25
36′ span, 16″ O.C.	.022	Ea.	.021	2.42	.82	3.24
24″ O.C.	.015	Ea.	.015	1.65	.56	2.21

Hip Roof Framing Systems

System Description	QUAN.	UNIT	LABOR-HOURS	COST PER S.F.			S.F. EXTENSIONS
				MAT.	INST.	TOTAL	
2″ X 6″, 16″ O.C., 4/12 PITCH							
Hip rafters, 2″ x 6″, 4/12 pitch	.160	L.F.	.003	.09	.09	.18	
Jack rafters, 2″ x 6″, 16″ O.C., 4/12 pitch	1.430	L.F.	.038	.83	1.01	1.84	
Ceiling joists, 2″ x 4″, 16″ O.C.	1.000	L.F.	.013	.40	.35	.75	
Fascia board, 2″ x 6″	.220	L.F.	.012	.14	.33	.47	
Soffit nailer (outrigger), 2″ x 4″, 24″ O.C.	.220	L.F.	.006	.09	.16	.25	
Sheathing, 1/2″ exterior plywood, CDX	1.570	S.F.	.018	.79	.48	1.27	
Furring strips, 1″ x 3″, 16″ O.C.	1.000	L.F.	.023	.35	.67	1.02	
TOTAL			.113	2.69	3.09	5.78	
2″ X 6″, 24″ O.C., 4/12 PITCH							
Hip rafters, 2″ x 6″, 4/12 pitch	.160	L.F.	.003	.09	.09	.18	
Jack rafters, 2″ x 6″, 24″ O.C., 4/12 pitch	1.150	L.F.	.031	.67	.81	1.48	
Ceiling joists, 2″ x 4″, 24″ O.C.	.750	L.F.	.010	.30	.27	.57	
Fascia board, 2″ x 6″	.220	L.F.	.016	.18	.42	.60	
Soffit nailer (outrigger), 2″ x 4″, 24″ O.C.	.220	L.F.	.006	.09	.16	.25	
Sheathing, 1/2″ exterior plywood, CDX	1.570	S.F.	.018	.79	.48	1.27	
Furring strips, 1″ x 3″, 16″ O.C.	1.000	L.F.	.023	.35	.67	1.02	
TOTAL			.107	2.47	2.90	5.37	
2″ X 8″, 16″ O.C., 4/12 PITCH							
Hip rafters, 2″ x 8″, 4/12 pitch	.160	L.F.	.004	.13	.09	.22	
Jack rafters, 2″ x 8″, 16″ O.C., 4/12 pitch	1.430	L.F.	.047	1.19	1.26	2.45	
Ceiling joists, 2″ x 6″, 16″ O.C.	1.000	L.F.	.013	.58	.35	.93	
Fascia board, 2″ x 8″	.220	L.F.	.016	.18	.42	.60	
Soffit nailer (outrigger), 2″ x 4″, 24″ O.C.	.220	L.F.	.006	.09	.16	.25	
Sheathing, 1/2″ exterior plywood, CDX	1.570	S.F.	.018	.79	.48	1.27	
Furring strips, 1″ x 3″, 16″ O.C.	1.000	L.F.	.023	.35	.67	1.02	
TOTAL			.127	3.31	3.43	6.74	

Hip Roof Framing Systems

System Description	QUAN.	UNIT	LABOR-HOURS	COST PER S.F. MAT.	COST PER S.F. INST.	COST PER S.F. TOTAL	S.F. EXTENSIONS
2" X 8", 24" O.C., 4/12 PITCH							
Hip rafters, 2" x 8", 4/12 pitch	.160	L.F.	.003	.09	.09	.18	
Jack rafters, 2" x 8", 24" O.C., 4/12 pitch	1.150	L.F.	.038	.95	1.02	1.97	
Ceiling joists, 2" x 6", 24" O.C.	.750	L.F.	.010	.44	.27	.71	
Fascia board, 2" x 8"	.220	L.F.	.012	.14	.33	.47	
Soffit nailer (outrigger), 2" x 4", 24" O.C.	.220	L.F.	.006	.09	.16	.25	
Sheathing, 1/2" exterior plywood, CDX	1.570	S.F.	.018	.79	.48	1.27	
Furring strips, 1" x 3", 16" O.C.	1.000	L.F.	.023	.35	.67	1.02	
TOTAL			.110	2.85	3.02	5.87	
2" X 10", 16" O.C., 4/12 PITCH							
Hip rafters, 2" x 10", 4/12 pitch	.160	L.F.	.004	.19	.13	.32	
Jack rafters, 2" x 10", 16" O.C., 4/12 pitch	1.430	L.F.	.051	1.72	1.37	3.09	
Ceiling joists, 2" x 8", 16" O.C.	1.000	L.F.	.015	.83	.39	1.22	
Fascia board, 2" x 10"	.220	L.F.	.012	.14	.33	.47	
Soffit nailer (outrigger), 2" x 4", 24" O.C.	.220	L.F.	.006	.09	.16	.25	
Sheathing, 1/2" exterior plywood, CDX	1.570	S.F.	.018	.79	.48	1.27	
Furring strips, 1" x 3", 16" O.C.	1.000	L.F.	.023	.35	.67	1.02	
TOTAL			.129	4.11	3.53	7.64	
2" X 10", 24" O.C., 4/12 PITCH							
Hip rafters, 2" x 10", 4/12 pitch	.160	L.F.	.004	.19	.13	.32	
Jack rafters, 2" x 8", 24" O.C., 4/12 pitch	1.150	L.F.	.038	.95	1.02	1.97	
Ceiling joists, 2" x 8", 24" O.C.	.750	L.F.	.011	.62	.30	.92	
Fascia board, 2" x 10"	.220	L.F.	.012	.14	.33	.47	
Soffit nailer (outrigger), 2" x 4", 24" O.C.	.220	L.F.	.006	.09	.16	.25	
Sheathing, 1/2" exterior plywood, CDX	1.570	S.F.	.018	.79	.48	1.27	
Furring strips, 1" x 3", 16" O.C.	1.000	L.F.	.023	.35	.67	1.02	
TOTAL			.112	3.13	3.09	6.22	

The cost of this system is based on S.F. of plan area. Measurement is area under the hip roof only. See gable roof system for added costs.

Hip Roof Framing System Components

Component Description	QUAN.	UNIT	LABOR-HOURS	COST PER S.F. MAT.	COST PER S.F. INST.	COST PER S.F. TOTAL
Jack rafters						
Jack rafters, #2 or better, 16" O.C., 2" x 6", 4/12 pitch	1.430	L.F.	.038	.83	1.01	1.84
8/12 pitch	1.800	L.F.	.061	1.04	1.62	2.66
2" x 8", 4/12 pitch	1.430	L.F.	.047	1.19	1.26	2.45
8/12 pitch	1.800	L.F.	.075	1.49	2.02	3.51
24" O.C., 2" x 6", 4/12 pitch	1.150	L.F.	.031	.67	.81	1.48
8/12 pitch	1.440	L.F.	.049	.84	1.30	2.14
2" x 8", 4/12 pitch	1.150	L.F.	.038	.95	1.02	1.97
8/12 pitch	1.440	L.F.	.066	1.73	1.77	3.50
Ceiling Joists						
2" x 10", 16" O.C.	1.000	L.F.	.018	1.20	.48	1.68
24" O.C.	.750	L.F.	.013	.90	.37	1.27
Sheathing						
Sheathing, plywood CDX, 4/12 pitch, 3/8" thick	1.570	S.F.	.016	.63	.44	1.07
5/8" thick	1.570	S.F.	.019	.91	.53	1.44

Gambrel Roof Framing Systems

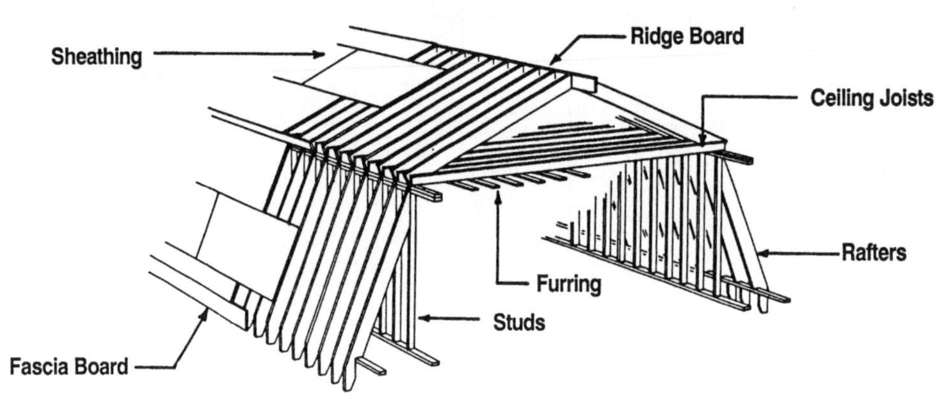

System Description	QUAN.	UNIT	LABOR-HOURS	COST PER S.F.			S.F. EXTENSIONS
				MAT.	INST.	TOTAL	
2'' X 6'' RAFTERS, 16'' O.C.							
Roof rafters, 2'' x 6'', 16'' O.C.	1.430	L.F.	.029	.83	.77	1.60	
Ceiling joists, 2'' x 6'', 16'' O.C.	.710	L.F.	.009	.41	.24	.65	
Stud wall, 2'' x 4'', 16'' O.C., including plates	.790	L.F.	.012	.32	.34	.66	
Furring strips, 1'' x 3'', 16'' O.C.	.710	L.F.	.016	.25	.48	.73	
Ridge board, 2'' x 8''	.050	L.F.	.002	.04	.05	.09	
Fascia board, 2'' x 6''	.100	L.F.	.006	.07	.16	.23	
Sheathing, exterior grade plywood, 1/2'' thick	1.450	S.F.	.017	.73	.43	1.16	
TOTAL			.091	2.65	2.47	5.12	
2'' X 6'' RAFTERS, 24'' O.C.							
Roof rafters, 2'' x 6'', 24'' O.C.	1.140	L.F.	.023	.66	.61	1.27	
Ceiling joists, 2'' x 6'', 24'' O.C.	.570	L.F.	.007	.33	.20	.53	
Stud wall, 2'' x 4'', 24'' O.C., including plates	.630	L.F.	.010	.25	.26	.51	
Furring strips, 1'' x 3'', 16'' O.C.	.710	L.F.	.016	.25	.48	.73	
Ridge board, 2'' x 8''	.050	L.F.	.002	.04	.05	.09	
Fascia board, 2'' x 6''	.100	L.F.	.006	.07	.16	.23	
Sheathing, exterior grade plywood, 1/2'' thick	1.450	S.F.	.017	.73	.43	1.16	
TOTAL			.081	2.33	2.19	4.52	
2'' X 8'' RAFTERS, 16'' O.C.							
Roof rafters, 2'' x 8'', 16'' O.C.	1.430	L.F.	.031	1.19	.83	2.02	
Ceiling joists, 2'' x 6'', 16'' O.C.	.710	L.F.	.009	.41	.24	.65	
Stud wall, 2'' x 4'', 16'' O.C., including plates	.790	L.F.	.012	.32	.34	.66	
Furring strips, 1'' x 3'', 16'' O.C.	.710	L.F.	.016	.25	.48	.73	
Ridge board, 2'' x 8''	.050	L.F.	.002	.04	.05	.09	
Fascia board, 2'' x 8''	.100	L.F.	.007	.08	.19	.27	
Sheathing, exterior grade plywood, 1/2'' thick	1.450	S.F.	.017	.73	.43	1.16	
TOTAL			.094	3.02	2.56	5.58	

Gambrel Roof Framing Systems

System Description	QUAN.	UNIT	LABOR-HOURS	MAT.	INST.	TOTAL	S.F. EXTENSIONS
2" X 8" RAFTERS, 24" O.C.							
Roof rafters, 2" x 8", 24" O.C.	1.140	L.F.	.024	.95	.66	1.61	
Ceiling joists, 2" x 6", 24" O.C.	.570	L.F.	.007	.33	.20	.53	
Stud wall, 2" x 4", 24" O.C., including plates	.630	L.F.	.010	.25	.26	.51	
Furring strips, 1" x 3", 16" O.C.	.710	L.F.	.016	.25	.48	.73	
Ridge board, 2" x 8"	.050	L.F.	.002	.04	.05	.09	
Fascia board, 2" x 8"	.100	L.F.	.007	.08	.19	.27	
Sheathing, exterior grade plywood, 1/2" thick	1.450	S.F.	.017	.73	.43	1.16	
TOTAL			.083	2.63	2.27	4.90	
2" X 10" RAFTERS, 16" O.C.							
Roof rafters, 2" x 10", 16" O.C.	1.430	L.F.	.046	1.72	1.25	2.97	
Ceiling joists, 2" x 8", 16" O.C.	.710	L.F.	.010	.59	.27	.86	
Stud wall, 2" x 6", 16" O.C., including plates	.790	L.F.	.014	.46	.38	.84	
Furring strips, 1" x 3", 16" O.C.	.710	L.F.	.016	.25	.48	.73	
Ridge board, 2" x 10"	.050	L.F.	.002	.06	.05	.11	
Fascia board, 2" x 10"	.100	L.F.	.009	.12	.24	.36	
Sheathing, exterior grade plywood, 1/2" thick	1.450	S.F.	.017	.73	.43	1.16	
TOTAL			.114	3.93	3.10	7.03	
2" X 10" RAFTERS, 24" O.C.							
Roof rafters, 2" x 10", 24" O.C.	1.140	L.F.	.037	1.37	1.00	2.37	
Ceiling joists, 2" x 8", 24" O.C.	.570	L.F.	.008	.47	.22	.69	
Stud wall, 2" x 6", 24" O.C., including plates	.630	L.F.	.011	.37	.30	.67	
Furring strips, 1" x 3", 16" O.C.	.710	L.F.	.016	.25	.48	.73	
Ridge board, 2" x 10"	.050	L.F.	.002	.06	.05	.11	
Fascia board, 2" x 10"	.100	L.F.	.009	.12	.24	.36	
Sheathing, exterior grade plywood, 1/2" thick	1.450	S.F.	.017	.73	.43	1.16	
TOTAL			.100	3.37	2.72	6.09	

The cost of this system is based on the square foot of plan area on the first floor.

Gambrel Roof Framing System Components

Component Description	QUAN.	UNIT	LABOR-HOURS	MAT.	INST.	TOTAL
Ceiling joists						
Ceiling joist, #2 or better, 2" x 4", 16" O.C.	.710	L.F.	.009	.28	.24	.52
24" O.C.	.570	L.F.	.007	.23	.20	.43
Furring						
Furring, 1" x 3", 16" O.C.	.710	L.F.	.016	.25	.48	.73
24" O.C.	.590	L.F.	.013	.21	.40	.61
Ridge board						
Ridge board, #2 or better, 1" x 6"	.050	L.F.	.001	.03	.03	.06
1" x 8"	.050	L.F.	.001	.04	.04	.08
1" x 10"	.050	L.F.	.002	.05	.04	.09
2" x 6"	.050	L.F.	.002	.03	.04	.07
Fascia board						
Fascia board, #2 or better, 1" x 6"	.100	L.F.	.004	.05	.10	.15
1" x 8"	.100	L.F.	.005	.05	.13	.18
1" x 10"	.100	L.F.	.005	.06	.14	.20

Mansard Roof Framing Systems

System Description	QUAN.	UNIT	LABOR-HOURS	COST PER S.F.			S.F. EXTENSIONS
				MAT.	INST.	TOTAL	
2″ X 6″ RAFTERS, 16″ O.C.							
Roof rafters, 2″ x 6″, 16″ O.C.	1.210	L.F.	.033	.70	.89	1.59	
Rafter plates, 2″ x 6″, double top, single bottom	.364	L.F.	.010	.21	.26	.47	
Ceiling joists, 2″ x 4″, 16″ O.C.	.920	L.F.	.012	.37	.32	.69	
Hip rafter, 2″ x 6″	.070	L.F.	.002	.04	.06	.10	
Jack rafter, 2″ x 6″, 16″ O.C.	1.000	L.F.	.039	.58	1.05	1.63	
Ridge board, 2″ x 6″	.018	L.F.	.001	.01	.01	.02	
Sheathing, exterior grade plywood, 1/2″ thick	2.210	S.F.	.025	1.11	.66	1.77	
Furring strips, 1″ x 3″, 16″ O.C.	.920	L.F.	.021	.32	.62	.94	
TOTAL			.143	3.34	3.87	7.21	
2″ X 6″ RAFTERS, 24″ O.C.							
Roof rafters, 2″ x 6″, 24″ O.C.	.970	L.F.	.026	.56	.71	1.27	
Rafter plates, 2″ x 6″, double top, single bottom	.364	L.F.	.010	.21	.26	.47	
Ceiling joists, 2″ x 4″, 24″ O.C.	.740	L.F.	.009	.30	.25	.55	
Hip rafter, 2″ x 6″	.070	L.F.	.002	.04	.06	.10	
Jack rafter, 2″ x 6″, 24″ O.C.	.800	L.F.	.031	.46	.84	1.30	
Ridge board, 2″ x 6″	.018	L.F.	.001	.01	.01	.02	
Sheathing, exterior grade plywood, 1/2″ thick	2.210	S.F.	.025	1.11	.66	1.77	
Furring strips, 1″ x 3″, 16″ O.C.	.920	L.F.	.021	.32	.62	.94	
TOTAL			.125	3.01	3.41	6.42	
2″ X 8″ RAFTERS, 16″ O.C.							
Roof rafters, 2″ x 8″, 16″ O.C.	1.210	L.F.	.036	1.00	.97	1.97	
Rafter plates, 2″ x 8″, double top, single bottom	.364	L.F.	.011	.30	.29	.59	
Ceiling joists, 2″ x 6″, 16″ O.C.	.920	L.F.	.012	.53	.32	.85	
Hip rafter, 2″ x 8″	.070	L.F.	.002	.06	.06	.12	
Jack rafter, 2″ x 8″, 16″ O.C.	1.000	L.F.	.048	.83	1.28	2.11	
Ridge board, 2″ x 8″	.018	L.F.	.001	.01	.02	.03	
Sheathing, exterior grade plywood, 1/2″ thick	2.210	S.F.	.025	1.11	.66	1.77	
Furring strips, 1″ x 3″, 16″ O.C.	.920	L.F.	.021	.32	.62	.94	
TOTAL			.156	4.16	4.22	8.38	

Mansard Roof Framing Systems

System Description	QUAN.	UNIT	LABOR-HOURS	COST PER S.F.			S.F. EXTENSIONS
				MAT.	INST.	TOTAL	
2″ X 8″ RAFTERS, 24″ O.C.							
Roof rafters, 2″ x 8″, 24″ O.C.	.970	L.F.	.029	.81	.78	1.59	
Rafter plates, 2″ x 8″, double top, single bottom	.364	L.F.	.011	.30	.29	.59	
Ceiling joists, 2″ x 6″, 24″ O.C.	.740	L.F.	.011	.61	.28	.89	
Hip rafter, 2″ x 8″	.070	L.F.	.002	.06	.06	.12	
Jack rafter, 2″ x 8″, 24″ O.C.	.800	L.F.	.038	.66	1.03	1.69	
Ridge board, 2″ x 8″	.018	L.F.	.001	.01	.02	.03	
Sheathing, exterior grade plywood, 1/2″ thick	2.210	S.F.	.025	1.11	.66	1.77	
Furring strips, 1″ x 3″, 16″ O.C.	.920	L.F.	.021	.32	.62	.94	
TOTAL			.138	3.88	3.74	7.62	
2″ X 10″ RAFTERS, 16″ O.C.							
Roof rafters, 2″ x 10″, 16″ O.C.	1.210	L.F.	.046	1.45	1.22	2.67	
Rafter plates, 2″ x 10″, double top, single bottom	.364	L.F.	.014	.44	.37	.81	
Ceiling joists, 2″ x 8″, 16″ O.C.	.920	L.F.	.013	.76	.36	1.12	
Hip rafter, 2″ x 10″	.070	L.F.	.003	.08	.08	.16	
Jack rafter, 2″ x 8″, 16″ O.C.	1.000	L.F.	.048	.83	1.28	2.11	
Ridge board, 2″ x 10″	.018	L.F.	.001	.02	.02	.04	
Sheathing, exterior grade plywood, 1/2″ thick	2.210	S.F.	.025	1.11	.66	1.77	
Furring strips, 1″ x 3″, 16″ O.C.	.920	L.F.	.021	.32	.62	.94	
TOTAL			.171	5.01	4.61	9.62	
2″ X 10″ RAFTERS, 24″ O.C.							
Roof rafters, 2″ x 10″, 24″ O.C.	.970	L.F.	.037	1.16	.98	2.14	
Rafter plates, 2″ x 10″, double top, single bottom	.364	L.F.	.014	.44	.37	.81	
Ceiling joists, 2″ x 8″, 24″ O.C.	.740	L.F.	.011	.61	.28	.89	
Hip rafter, 2″ x 10″	.070	L.F.	.003	.08	.08	.16	
Jack rafter, 2″ x 8″, 24″ O.C.	.800	L.F.	.038	.66	1.03	1.69	
Ridge board, 2″ x 10″	.018	L.F.	.001	.02	.02	.04	
Sheathing, exterior grade plywood, 1/2″ thick	2.210	S.F.	.025	1.11	.66	1.77	
Furring strips, 1″ x 3″, 16″ O.C.	.920	L.F.	.021	.32	.62	.94	
TOTAL			.150	4.40	4.04	8.44	

The cost of this system is based on the square foot of plan area.

Mansard Roof Framing System Components

Component Description	QUAN.	UNIT	LABOR-HOURS	COST PER S.F.		
				MAT.	INST.	TOTAL
Ridge board						
Ridge board, #2 or better, 1″ x 6″	.018	L.F.	.001	.01	.01	.02
1″ x 8″	.018	L.F.	.001	.01	.01	.02
Sheathing						
Sheathing, plywood exterior grade CDX, 3/8″ thick	2.210	S.F.	.023	.88	.62	1.50
1/2″ thick	2.210	S.F.	.025	1.11	.66	1.77
5/8″ thick	2.210	S.F.	.027	1.28	.75	2.03
3/4″ thick	2.210	S.F.	.029	1.64	.79	2.43
Boards, 1″ x 6″, laid regular	2.210	S.F.	.049	2.41	1.33	3.74
Laid diagonal	2.210	S.F.	.054	2.41	1.46	3.87
1″ x 8″, laid regular	2.210	S.F.	.040	2.50	1.08	3.58
Laid diagonal	2.210	S.F.	.049	2.50	1.33	3.83

Shed/Flat Roof Framing Systems

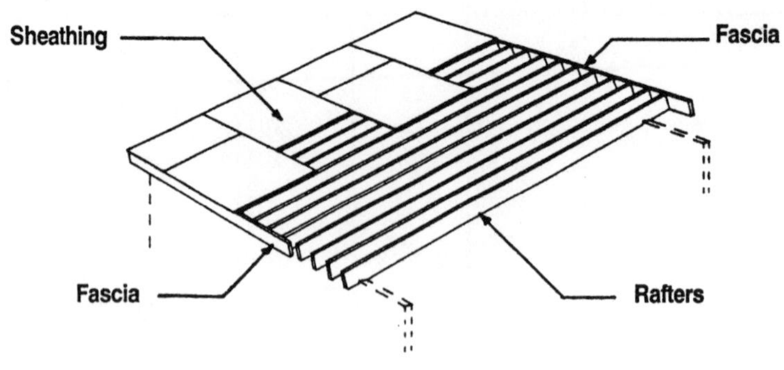

System Description	QUAN.	UNIT	LABOR-HOURS	COST PER S.F.			S.F. EXTENSIONS
				MAT.	INST.	TOTAL	
2″ X 4″,16″ O.C., 4/12 PITCH							
Rafters, 2″ x 4″, 16″ O.C., 4/12 pitch	1.170	L.F.	.014	.50	.38	.88	
Fascia, 2″ x 4″	.100	L.F.	.005	.06	.13	.19	
Bridging, 1″ x 3″, 6′ OC	.080	Pr.	.005	.04	.15	.19	
Sheathing, exterior grade plywood, 1/2″ thick	1.230	S.F.	.014	.62	.37	.99	
TOTAL			.038	1.22	1.03	2.25	
2″ X 4″, 24″ O.C., 4/12 PITCH							
Rafters, 2″ x 4″, 24″ O.C., 4/12 pitch	.940	L.F.	.011	.41	.30	.71	
Fascia, 2″ x 4″	.100	L.F.	.005	.06	.13	.19	
Bridging, 1″ x 3″, 6′ OC	.060	Pr.	.005	.04	.15	.19	
Sheathing, exterior grade plywood, 1/2″ thick	1.230	S.F.	.014	.62	.37	.99	
TOTAL			.035	1.13	.95	2.08	
2″ X 6″,16″ O.C., 4/12 PITCH							
Rafters, 2″ x 6″, 16″ O.C., 4/12 pitch	1.170	L.F.	.019	.68	.50	1.18	
Fascia, 2″ x 6″	.100	L.F.	.006	.07	.16	.23	
Bridging, 1″ x 3″, 6′ O.C.	.080	Pr.	.005	.04	.15	.19	
Sheathing, exterior grade plywood, 1/2″ thick	1.230	S.F.	.014	.62	.37	.99	
TOTAL			.044	1.41	1.18	2.59	
2″ X 6″, 24″ O.C., 4/12 PITCH							
Rafters, 2″ x 6″, 24″ O.C., 4/12 pitch	.940	L.F.	.015	.55	.41	.96	
Fascia, 2″ x 6″	.100	L.F.	.006	.07	.16	.23	
Bridging, 1″ x 3″, 6′ O.C.	.060	Pr.	.004	.03	.11	.14	
Sheathing, exterior grade plywood, 1/2″ thick	1.230	S.F.	.014	.62	.37	.99	
TOTAL			.039	1.27	1.05	2.32	
2″ X 8″, 16″ O.C., 4/12 PITCH							
Rafters, 2″ x 8″, 16″ O.C., 4/12 pitch	1.170	L.F.	.020	.97	.52	1.49	
Fascia, 2″ x 8″	.100	L.F.	.007	.08	.19	.27	
Bridging, 1″ x 3″, 6′ O.C.	.080	Pr.	.005	.04	.15	.19	
Sheathing, exterior grade plywood, 1/2″ thick	1.230	S.F.	.014	.62	.37	.99	
TOTAL			.046	1.71	1.23	2.94	

Shed/Flat Roof Framing Systems

System Description	QUAN.	UNIT	LABOR-HOURS	COST PER S.F.			S.F. EXTENSIONS
				MAT.	INST.	TOTAL	
2″ X 8″, 24″ O.C., 4/12 PITCH							
Rafters, 2″ x 8″, 24″ O.C., 4/12 pitch	.940	L.F.	.016	.78	.42	1.20	
Fascia, 2″ x 8″	.100	L.F.	.007	.08	.19	.27	
Bridging, 1″ x 3″, 6′ O.C.	.060	Pr.	.004	.03	.11	.14	
Sheathing, exterior grade plywood, 1/2″ thick	1.230	S.F.	.014	.62	.37	.99	
TOTAL			.041	1.51	1.09	2.60	
2″ X 10″, 16″ O.C., 4/12 PITCH							
Rafters, 2″ x 10″, 16″ O.C., 4/12 pitch	1.170	L.F.	.030	1.40	.80	2.20	
Fascia, 2″ x 10″	.100	L.F.	.009	.12	.24	.36	
Bridging, 1″ x 3″, 6′ OC	.080	Pr.	.004	.03	.11	.14	
Sheathing, exterior grade plywood, 1/2″ thick	1.230	S.F.	.014	.62	.37	.99	
TOTAL			.057	2.17	1.52	3.69	
2″ X 10″, 24″ O.C., 4/12 PITCH							
Rafters, 2″ x 10″, 24″ O.C., 4/12 pitch	.940	L.F.	.024	1.13	.64	1.77	
Fascia, 2″ x 10″	.100	L.F.	.009	.12	.24	.36	
Bridging, 1″ x 3″, 6′ OC	.060	Pr.	.004	.03	.11	.14	
Sheathing, exterior grade plywood, 1/2″ thick	1.230	S.F.	.014	.62	.37	.99	
TOTAL			.051	1.90	1.36	3.26	

The cost of this system is based on the square foot of plan area.
A 1′ overhang is assumed. No ceiling joists or furring are included.

Shed/Flat Roof Framing System Components

Component Description	QUAN.	UNIT	LABOR-HOURS	COST PER S.F.		
				MAT.	INST.	TOTAL
Rafters						
Rafters, #2 or better, 16″ O.C., 2″ x 4″, 0 - 4/12 pitch	1.170	L.F.	.014	.50	.38	.88
2″ x 8″	1.330	L.F.	.028	1.10	.77	1.87
24″ O.C., 2″ x 4″	1.060	L.F.	.021	.61	.57	1.18
2″ x 6″	1.060	L.F.	.021	.61	.57	1.18
2″ x 8″	1.060	L.F.	.023	.88	.61	1.49
2″ x 10″	1.060	L.F.	.034	1.27	.92	2.19
Fascia						
Fascia, #2 or better,, 1″ x 4″	.100	L.F.	.003	.03	.07	.10
1″ x 6″	.100	L.F.	.004	.05	.10	.15
Bridging						
Metal, galvanized, rafters, 16″ O.C.	.080	Pr.	.005	.08	.14	.22
24″ O.C.	.060	Pr.	.003	.08	.10	.18
Compression type, rafters, 16″ O.C.	.080	Pr.	.003	.10	.09	.19
24″ O.C.	.060	Pr.	.002	.08	.07	.15
Sheathing						
Sheathing, plywood, exterior grade, 3/8″ thick, flat 0 - 4/12 pitch	1.230	S.F.	.013	.49	.34	.83
5/12 - 8/12 pitch	1.330	S.F.	.014	.53	.37	.90
1/2″ thick, flat 0 - 4/12 pitch	1.230	S.F.	.014	.62	.37	.99
5/12 - 8/12 pitch	1.330	S.F.	.015	.67	.40	1.07
5/8″ thick, flat 0 - 4/12 pitch	1.230	S.F.	.015	.71	.41	1.12
5/12 - 8/12 pitch	1.330	S.F.	.016	.77	.46	1.23

Gable Dormer Framing Systems

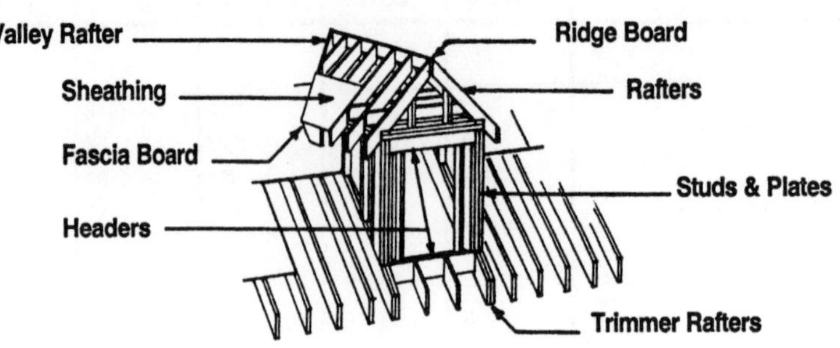

System Description	QUAN.	UNIT	LABOR-HOURS	COST PER S.F.			S.F. EXTENSIONS
				MAT.	INST.	TOTAL	
2" x 4", 16" O.C.							
Dormer rafter, 2" x 4", 16" O.C.	1.330	L.F.	.029	.62	.77	1.39	
Ridge board, 2" x 4"	.280	L.F.	.007	.13	.19	.32	
Trimmer rafters, 2" x 6"	.880	L.F.	.014	.51	.38	.89	
Wall studs & plates, 2" x 4", 16" O.C.	3.160	L.F.	.056	1.26	1.48	2.74	
Fascia, 2" x 4"	.220	L.F.	.011	.12	.28	.40	
Valley rafter, 2" x 4", 16" O.C.	.280	L.F.	.007	.13	.19	.32	
Cripple rafter, 2" x 4", 16" O.C.	.560	L.F.	.018	.26	.47	.73	
Headers, 2" x 4", doubled	.670	L.F.	.024	.31	.65	.96	
Ceiling joist, 2" x 4", 16" O.C.	1.000	L.F.	.013	.40	.35	.75	
Sheathing, exterior grade plywood, 1/2" thick	3.610	S.F.	.041	1.81	1.09	2.90	
TOTAL			.220	5.55	5.85	11.40	
2" X 6", 16" O.C.							
Dormer rafter, 2" x 6", 16" O.C.	1.330	L.F.	.036	.77	.97	1.74	
Ridge board, 2" x 6"	.280	L.F.	.009	.16	.24	.40	
Trimmer rafters, 2" x 6"	.880	L.F.	.014	.51	.38	.89	
Wall studs & plates, 2" x 4", 16" O.C.	3.160	L.F.	.056	1.26	1.48	2.74	
Fascia, 2" x 6"	.220	L.F.	.012	.14	.33	.47	
Valley rafter, 2" x 6", 16" O.C.	.280	L.F.	.009	.16	.23	.39	
Cripple rafter, 2" x 6", 16" O.C.	.560	L.F.	.022	.32	.59	.91	
Headers, 2" x 6", doubled	.670	L.F.	.030	.39	.80	1.19	
Ceiling joist, 2" x 4", 16" O.C.	1.000	L.F.	.013	.40	.35	.75	
Sheathing, exterior grade plywood, 1/2" thick	3.610	S.F.	.041	1.81	1.09	2.90	
TOTAL			.242	5.92	6.46	12.38	
2" x 6", 24" O.C.							
Dormer rafter, 2" x 6", 24" O.C.	1.060	L.F.	.029	.61	.77	1.38	
Ridge board, 2" x 6"	.280	L.F.	.009	.16	.24	.40	
Trimmer rafters, 2" x 6"	.880	L.F.	.014	.51	.38	.89	
Wall studs & plates, 2" x 4", 24" O.C.	2.800	L.F.	.050	1.12	1.32	2.44	
Fascia, 2" x 6"	.220	L.F.	.012	.14	.33	.47	
Valley rafter, 2" x 6", 24" O.C.	.280	L.F.	.009	.16	.23	.39	
Cripple rafter, 2" x 6", 24" O.C.	.450	L.F.	.018	.26	.47	.73	
Headers, 2" x 6", doubled	.670	L.F.	.030	.39	.80	1.19	
Ceiling joist, 2" x 4", 24" O.C.	.800	L.F.	.010	.32	.28	.60	
Sheathing, exterior grade plywood, 1/2" thick	3.610	S.F.	.041	1.81	1.09	2.90	
TOTAL			.222	5.48	5.91	11.39	

Gable Dormer Framing Systems

System Description	QUAN.	UNIT	LABOR-HOURS	MAT.	INST.	TOTAL	S.F. EXTENSIONS
2" X 8", 16" O.C.							
Dormer rafter, 2" x 8", 16" O.C.	1.330	L.F.	.039	1.10	1.06	2.16	
Ridge board, 2" x 8"	.280	L.F.	.010	.23	.26	.49	
Trimmer rafter, 2" x 8"	.880	L.F.	.015	.73	.40	1.13	
Wall studs & plates, 2" x 4", 16" O.C.	3.160	L.F.	.056	1.26	1.48	2.74	
Fascia, 2" x 8"	.220	L.F.	.016	.18	.42	.60	
Valley rafter, 2" x 8", 16" O.C.	.280	L.F.	.010	.23	.25	.48	
Cripple rafter, 2" x 8", 16" O.C.	.560	L.F.	.027	.46	.71	1.17	
Headers, 2" x 8", doubled	.670	L.F.	.032	.56	.85	1.41	
Ceiling joist, 2" x 4", 16" O.C.	1.000	L.F.	.013	.40	.35	.75	
Sheathing,, exterior grade plywood, 1/2" thick	3.610	S.F.	.041	1.81	1.09	2.90	
TOTAL			.259	6.96	6.87	13.83	
2" x 8", 24" O.C.							
Dormer rafter, 2" x 8", 24" O.C.	1.060	L.F.	.031	.88	.85	1.73	
Ridge board, 2" x 8"	.280	L.F.	.010	.23	.26	.49	
Trimmer rafter, 2" x 8"	.880	L.F.	.015	.73	.40	1.13	
Wall studs & plates, 2" x 6", 24" O.C.	2.800	L.F.	.056	1.62	1.51	3.13	
Fascia, 2" x 8"	.220	L.F.	.016	.18	.42	.60	
Valley rafter, 2" x 8", 24" O.C.	.280	L.F.	.010	.23	.25	.48	
Cripple rafter, 2" x 8", 24" O.C.	.450	L.F.	.021	.37	.58	.95	
Headers, 2" x 8", doubled	.670	L.F.	.032	.56	.85	1.41	
Ceiling joist, 2" x 6", 24" O.C.	.800	L.F.	.010	.46	.28	.74	
Sheathing, exterior grade plywood, 1/2" thick	3.610	S.F.	.041	1.81	1.09	2.90	
TOTAL			.242	7.07	6.49	13.56	

The cost in this system is based on the square foot of plan area.
The measurement being the plan area of the dormer only.

Gable Dormer Framing System Components

Component Description	QUAN.	UNIT	LABOR-HOURS	MAT.	INST.	TOTAL
Ridge board						
Ridge board, #2 or better, 1" x 4"	.280	L.F.	.006	.12	.16	.28
1" x 6"	.280	L.F.	.007	.15	.20	.35
1" x 8"	.280	L.F.	.008	.21	.22	.43
2" x 4"	.280	L.F.	.007	.13	.19	.32
2" x 6"	.280	L.F.	.009	.16	.24	.40
2" x 8"	.280	L.F.	.010	.23	.26	.49
Fascia						
Fascia, #2 or better, 1" x 4"	.220	L.F.	.006	.07	.17	.24
1" x 6"	.220	L.F.	.008	.09	.21	.30
1" x 8"	.220	L.F.	.009	.11	.24	.35
2" x 4"	.220	L.F.	.011	.12	.28	.40
Sheathing						
Sheathing, plywood exterior grade, 3/8" thick	3.610	S.F.	.038	1.44	1.01	2.45
5/8" thick	3.610	S.F.	.044	2.09	1.23	3.32
3/4" thick	3.610	S.F.	.048	2.67	1.30	3.97
Boards, 1" x 6", laid regular	3.610	S.F.	.089	3.93	2.38	6.31
Laid diagonal	3.610	S.F.	.099	3.93	2.64	6.57
1" x 8", laid regular	3.610	S.F.	.076	4.08	2.06	6.14

Shed Dormer Framing Systems

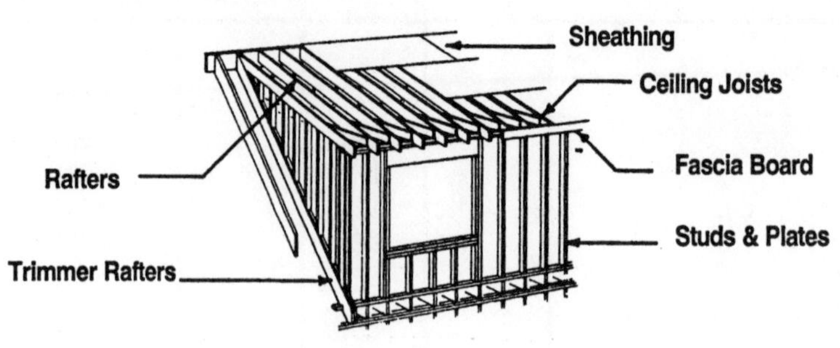

System Description	QUAN.	UNIT	LABOR-HOURS	COST PER S.F.			S.F. EXTENSIONS
				MAT.	INST.	TOTAL	
2" X 4" RAFTERS, 16" O.C.							
Dormer rafter, 2" x 4", 16" O.C.	1.080	L.F.	.023	.50	.63	1.13	
Trimmer rafter, 2" x 4"	.400	L.F.	.005	.19	.14	.33	
Studs & plates, 2" x 4", 16" O.C.	2.750	L.F.	.049	1.10	1.30	2.40	
Fascia, 2" x 4"	.250	L.F.	.011	.12	.28	.40	
Ceiling joist, 2" x 4", 16" O.C.	1.000	L.F.	.013	.40	.35	.75	
Sheathing, exterior grade plywood, CDX, 1/2" thick	2.940	S.F.	.034	1.47	.88	2.35	
TOTAL			.135	3.78	3.58	7.36	
2" X 6" RAFTERS, 16" O.C.							
Dormer rafter, 2" x 6", 16" O.C.	1.080	L.F.	.029	.63	.79	1.42	
Trimmer rafter, 2" x 6"	.400	L.F.	.006	.23	.17	.40	
Studs & plates, 2" x 4", 16" O.C.	2.750	L.F.	.049	1.10	1.30	2.40	
Fascia, 2" x 6"	.250	L.F.	.014	.16	.37	.53	
Ceiling joist, 2" x 4", 16" O.C.	1.000	L.F.	.013	.40	.35	.75	
Sheathing, exterior grade plywood, CDX, 1/2" thick	2.940	S.F.	.034	1.47	.88	2.35	
TOTAL			.145	3.99	3.86	7.85	
2" X 6" RAFTERS, 24" O.C.							
Dormer rafter, 2" x 6", 24" O.C.	.860	L.F.	.023	.50	.63	1.13	
Trimmer rafter, 2" x 6"	.400	L.F.	.006	.23	.17	.40	
Studs & plates, 2" x 4", 24" O.C.	2.750	L.F.	.049	1.10	1.30	2.40	
Fascia, 2" x 6"	.250	L.F.	.014	.16	.37	.53	
Ceiling joist, 2" x 4", 24" O.C.	.800	L.F.	.010	.32	.28	.60	
Sheathing, exterior grade plywood, CDX, 1/2" thick	2.940	S.F.	.034	1.47	.88	2.35	
TOTAL			.136	3.78	3.63	7.41	
2" X 8" RAFTERS, 16" O.C.							
Dormer rafter, 2" x 8", 16" O.C.	1.080	L.F.	.032	.90	.86	1.76	
Trimmer rafter, 2" x 8"	.400	L.F.	.007	.33	.18	.51	
Studs & plates, 2" x 4", 16" O.C.	2.750	L.F.	.049	1.10	1.30	2.40	
Fascia, 2" x 8"	.250	L.F.	.018	.21	.48	.69	
Ceiling joist, 2" x 6", 16" O.C.	1.000	L.F.	.013	.58	.35	.93	
Sheathing, exterior grade plywood, CDX, 1/2" thick	2.940	S.F.	.034	1.47	.88	2.35	
TOTAL			.153	4.59	4.05	8.64	

Shed Dormer Framing Systems

System Description	QUAN.	UNIT	LABOR-HOURS	COST PER S.F.			S.F. EXTENSIONS
				MAT.	INST.	TOTAL	
2″ X 8″ RAFTERS, 24″ O.C.							
Dormer rafter, 2″ x 8″, 24″ O.C.	.860	L.F.	.025	.71	.68	1.39	
Trimmer rafter, 2″ x 8″	.400	L.F.	.007	.33	.18	.51	
Studs & plates, 2″ x 4″, 24″ O.C.	2.750	L.F.	.055	1.60	1.48	3.08	
Fascia, 2″ x 8″	.250	L.F.	.018	.21	.48	.69	
Ceiling joist, 2″ x 6″, 24″ O.C.	.800	L.F.	.010	.46	.28	.74	
Sheathing, exterior grade plywood, CDX, 1/2″ thick	2.940	S.F.	.034	1.47	.88	2.35	
TOTAL			.149	4.78	3.98	8.76	
2″ X 10″ RAFTERS, 16″ O.C.							
Dormer rafter, 2″ x 10″, 16″ O.C.	1.080	L.F.	.041	1.30	1.09	2.39	
Trimmer rafter, 2″ x 10″	.400	L.F.	.010	.48	.27	.75	
Studs & plates, 2″ x 4″, 16″ O.C.	2.750	L.F.	.049	1.10	1.30	2.40	
Fascia, 2″ x 10″	.250	L.F.	.022	.30	.60	.90	
Ceiling joist, 2″ x 6″, 16″ O.C.	1.000	L.F.	.013	.58	.35	.93	
Sheathing, exterior grade plywood, CDX, 1/2″ thick	2.940	S.F.	.034	1.47	.88	2.35	
TOTAL			.169	5.23	4.49	9.72	
2″ X 10″ RAFTERS, 24″ O.C.							
Dormer rafter, 2″ x 10″, 24″ O.C.	.860	L.F.	.032	1.03	.87	1.90	
Trimmer rafter, 2″ x 10″	.400	L.F.	.010	.48	.27	.75	
Studs & plates, 2″ x 4″, 24″ O.C.	2.750	L.F.	.055	1.60	1.48	3.08	
Fascia, 2″ x 10″	.250	L.F.	.022	.30	.60	.90	
Ceiling joist, 2″ x 6″, 24″ O.C.	.800	L.F.	.010	.46	.28	.74	
Sheathing, exterior grade plywood, CDX, 1/2″ thick	2.940	S.F.	.034	1.47	.88	2.35	
TOTAL			.163	5.34	4.38	9.72	

The cost in this system is based on the square foot of plan area.
The measurement is the plan area of the dormer only.

Shed Dormer Framing System Components

Component Description	QUAN.	UNIT	LABOR-HOURS	COST PER S.F.		
				MAT.	INST.	TOTAL
Fascia boards						
Fascia, #2 or better, 1″ x 4″	.250	L.F.	.006	.07	.17	.24
1″ x 6″	.250	L.F.	.008	.09	.21	.30
Ceiling joists						
2″ x 8″, 16″ O.C.	1.000	L.F.	.015	.83	.39	1.22
24″ O.C.	.800	L.F.	.012	.66	.32	.98
Sheathing						
Sheathing, plywood exterior grade, 3/8″ thick	2.940	S.F.	.031	1.18	.82	2.00
5/8″ thick	2.940	S.F.	.036	1.71	1.00	2.71

Window Openings

Component Description	QUAN.	UNIT	LABOR-HOURS	COST EACH		
				MAT.	INST.	TOTAL
The following are to be added to the total cost of the dormers for window openings. Do not subtract window area from the stud wall quantities.						
Headers						
2″ x 6″ doubled, 2′ long	4.000	L.F.	.178	2.32	4.80	7.12
4′ long	8.000	L.F.	.356	4.64	9.60	14.24
2″ x 8″ doubled, 4′ long	8.000	L.F.	.376	6.65	10.20	16.85
8′ long	16.000	L.F.	.753	13.30	20.50	33.80

Partition Framing Systems

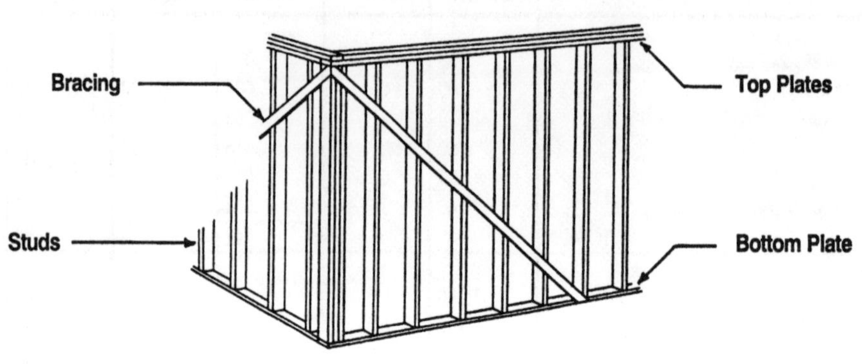

System Description	QUAN.	UNIT	LABOR-HOURS	COST PER S.F.			S.F. EXTENSIONS
				MAT.	INST.	TOTAL	
2" X 4", 12" O.C.							
2" x 4" studs, #2 or better, 12" O.C.	1.250	L.F.	.018	.50	.49	.99	
Plates, double top, single bottom	.375	L.F.	.005	.15	.15	.30	
Cross bracing, let-in, 1" x 6"	.080	L.F.	.004	.04	.14	.18	
TOTAL			.027	.69	.78	1.47	
2" X 4", 16" O.C.							
2" x 4" studs, #2 or better, 16" O.C.	1.000	L.F.	.015	.40	.39	.79	
Plates, double top, single bottom	.375	L.F.	.005	.15	.15	.30	
Cross bracing, let-in, 1" x 6"	.080	L.F.	.004	.04	.14	.18	
TOTAL			.024	.59	.68	1.27	
2" X 4", 24" O.C.							
2" x 4" studs, #2 or better, 24" O.C.	.800	L.F.	.012	.32	.32	.64	
Plates, double top, single bottom	.375	L.F.	.005	.15	.15	.30	
Cross bracing, let-in, 1" x 6"	.080	L.F.	.003	.04	.08	.12	
TOTAL			.020	.51	.55	1.06	
2" X 4", 32" O.C.							
2" x 4" studs, #2 or better, 32" O.C.	.650	L.F.	.009	.26	.25	.51	
Plates, double top, single bottom	.375	L.F.	.005	.15	.15	.30	
Cross bracing, let-in, 1" x 6"	.080	L.F.	.003	.04	.08	.12	
TOTAL			.017	.45	.48	.93	
2" X 6", 12" O.C.							
2" x 6" studs, #2 or better, 12" O.C.	1.250	L.F.	.020	.73	.54	1.27	
Plates, double top, single bottom	.375	L.F.	.006	.22	.16	.38	
Cross bracing, let-in, 1" x 6"	.080	L.F.	.004	.04	.14	.18	
TOTAL			.030	.99	.84	1.83	
2" X 6", 16" O.C.							
2" x 6" studs, #2 or better, 16" O.C.	1.000	L.F.	.016	.58	.43	1.01	
Plates, double top, single bottom	.375	L.F.	.006	.22	.16	.38	
Cross bracing, let-in, 1" x 6"	.080	L.F.	.004	.04	.14	.18	
TOTAL			.026	.84	.73	1.57	

Partition Framing Systems

System Description	QUAN.	UNIT	LABOR-HOURS	MAT.	INST.	TOTAL	S.F. EXTENSIONS
2″ X 6″, 24″ O.C.							
2″ x 6″ studs, #2 or better, 24″ O.C.	.800	L.F.	.013	.46	.35	.81	
Plates, double top, single bottom	.375	L.F.	.006	.22	.16	.38	
Cross bracing, let-in, 1″ x 6″	.080	L.F.	.003	.04	.08	.12	
TOTAL			.022	.72	.59	1.31	
2″ X 6″, 32″ O.C.							
2″ x 6″ studs, #2 or better, 32″ O.C.	.650	L.F.	.010	.38	.28	.66	
Plates, double top, single bottom	.375	L.F.	.006	.22	.16	.38	
Cross bracing, let-in, 1″ x 6″	.080	L.F.	.003	.04	.08	.12	
TOTAL			.019	.64	.52	1.16	

The costs in this system are based on a square foot of wall area. Do not subtract for door or window openings.

Partition Framing System Components

Component Description	QUAN.	UNIT	LABOR-HOURS	MAT.	INST.	TOTAL
Cross bracing						
Let-in steel (T shaped) studs, 12″ O.C.	.080	L.F.	.001	.06	.04	.10
16″ O.C.	.080	L.F.	.001	.05	.03	.08
24″ O.C.	.080	L.F.	.001	.05	.03	.08
32″ O.C.	.080	L.F.	.001	.04	.02	.06
Steel straps studs, 12″ O.C.	.080	L.F.	.001	.04	.03	.07
16″ O.C.	.080	L.F.	.001	.04	.03	.07
24″ O.C.	.080	L.F.	.001	.04	.03	.07
32″ O.C.	.080	L.F.	.001	.04	.03	.07
Metal studs						
Load bearing 24″ O.C., 20 ga. galv., 2-1/2″ wide	1.000	S.F.	.033	.72	.98	1.70
3-5/8″ wide	1.000	S.F.	.035	.84	1.02	1.86
4″ wide	1.000	S.F.	.036	.87	1.07	1.94
6″ wide	1.000	S.F.	.038	1.06	1.12	2.18
16 ga., 2-1/2″ wide	1.000	S.F.	.036	.91	1.07	1.98
3-5/8″ wide	1.000	S.F.	.038	1.01	1.12	2.13
4″ wide	1.000	S.F.	.040	1.10	1.17	2.27
6″ wide	1.000	S.F.	.042	1.35	1.24	2.59
Non-load bearing 24″ O.C., 25 ga. galv., 1-5/8″ wide	1.000	S.F.	.015	.20	.45	.65
2-1/2″ wide	1.000	S.F.	.016	.22	.46	.68
3-5/8″ wide	1.000	S.F.	.016	.27	.47	.74
4″ wide	1.000	S.F.	.016	.29	.48	.77
6″ wide	1.000	S.F.	.017	.38	.49	.87
20 ga., 2-1/2″ wide	1.000	S.F.	.016	.40	.46	.86
3-5/8″ wide	1.000	S.F.	.016	.47	.47	.94
4″ wide	1.000	S.F.	.016	.49	.48	.97
6″ wide	1.000	S.F.	.017	.62	.49	1.11

Exterior Closure Systems

Wood Siding Systems

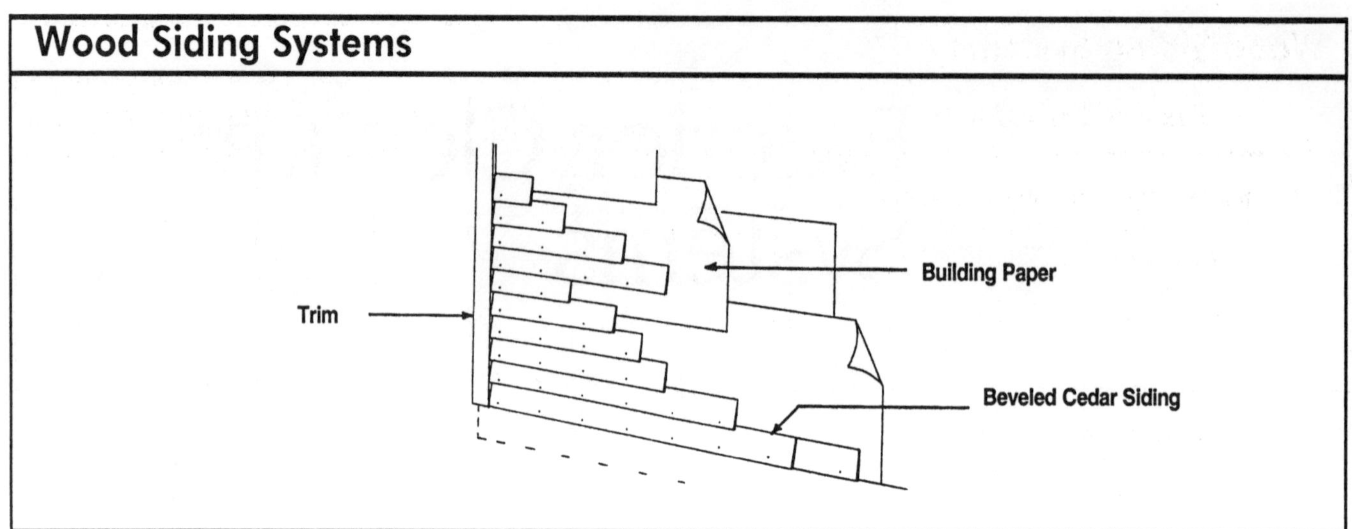

System Description	QUAN.	UNIT	LABOR-HOURS	COST PER S.F.			S.F. EXTENSIONS
				MAT.	INST.	TOTAL	
1/2" X 6" BEVELED CEDAR SIDING, "A" GRADE							
1/2" x 6" beveled cedar siding	1.000	S.F.	.032	2.22	.94	3.16	
#15 asphalt felt paper	1.100	S.F.	.002	.03	.07	.10	
Trim, cedar	.125	L.F.	.005	.17	.15	.32	
Paint, primer & 2 coats	1.000	S.F.	.017	.15	.45	.60	
TOTAL			.056	2.57	1.61	4.18	
1/2" X 8" BEVELED CEDAR SIDING, "A" GRADE							
1/2" x 8" beveled cedar siding	1.000	S.F.	.029	1.96	.85	2.81	
#15 asphalt felt paper	1.100	S.F.	.002	.03	.07	.10	
Trim, cedar	.125	L.F.	.005	.17	.15	.32	
Paint, primer & 2 coats	1.000	S.F.	.017	.15	.45	.60	
TOTAL			.053	2.31	1.52	3.83	
1" X 10" BOARD & BATTEN CEDAR SIDING, "B" GRADE							
1" x 10" board & batten cedar siding	1.000	S.F.	.031	2.09	.90	2.99	
#15 asphalt felt paper	1.100	S.F.	.002	.03	.07	.10	
Trim, cedar	.125	L.F.	.005	.17	.15	.32	
Paint, primer & 2 coats	1.000	S.F.	.017	.15	.45	.60	
TOTAL			.055	2.44	1.57	4.01	
1" X 10" BOARD & BATTEN WHITE PINE SIDING							
1" x 10" board & batten white pine siding	1.000	S.F.	.029	.73	.85	1.58	
#15 asphalt felt paper	1.100	S.F.	.002	.03	.07	.10	
Trim, pine	.125	L.F.	.005	.05	.15	.20	
Paint, primer & 2 coats	1.000	S.F.	.017	.15	.45	.60	
TOTAL			.053	.96	1.52	2.48	
1" X 4" TONGUE & GROOVE, REDWOOD, VERTICAL GRAIN							
1" x 4" tongue & groove, vertical, redwood	1.000	S.F.	.033	3.85	.98	4.83	
#15 asphalt felt paper	1.100	S.F.	.002	.03	.07	.10	
Trim, redwood	.125	L.F.	.005	.17	.15	.32	
Sealer, 1 coat, stain, 1 coat	1.000	S.F.	.013	.10	.35	.45	
TOTAL			.053	4.15	1.55	5.70	

Wood Siding Systems

System Description	QUAN.	UNIT	LABOR-HOURS	COST PER S.F.			S.F. EXTENSIONS
				MAT.	INST.	TOTAL	
1″ X 6″ TONGUE & GROOVE, REDWOOD, VERTICAL GRAIN							
1″ x 6″ tongue & groove, vertical, redwood	1.000	S.F.	.024	3.80	.69	4.49	
#15 asphalt felt paper	1.100	S.F.	.002	.03	.07	.10	
Trim, redwood	.125	L.F.	.005	.17	.15	.32	
Sealer, 1 coat, stain, 1 coat	1.000	S.F.	.013	.10	.35	.45	
TOTAL			.044	4.10	1.26	5.36	
3/8″ PLYWOOD SIDING, TEXTURE 1-11 CEDAR							
3/8″ plywood siding	1.000	S.F.	.024	1.19	.64	1.83	
#15 asphalt felt paper	1.100	S.F.	.002	.03	.07	.10	
Trim, cedar	.125	L.F.	.005	.17	.15	.32	
Paint, primer & 2 coats	1.000	S.F.	.017	.15	.45	.60	
TOTAL			.048	1.54	1.31	2.85	
3/8″ PLYWOOD SIDING, SOUTHERN YELLOW PINE							
3/8″ plywood siding	1.000	S.F.	.024	.62	.64	1.26	
#15 asphalt felt paper	1.100	S.F.	.002	.03	.07	.10	
Trim, pine	.125	L.F.	.005	.05	.15	.20	
Paint, primer & 2 coats	1.000	S.F.	.017	.15	.45	.60	
TOTAL			.048	.85	1.31	2.16	

The costs in this system are based on a square foot of wall area. Do not subtract area for door or window openings.

Wood Siding System Components

Component Description	QUAN.	UNIT	LABOR-HOURS	COST PER S.F.		
				MAT.	INST.	TOTAL
Siding						
Siding, beveled cedar, "B" grade, 1/2″ x 6″	1.000	S.F.	.028	1.02	.81	1.83
1/2″ x 8″	1.000	S.F.	.023	1.02	.67	1.69
Clear grade, 1/2″ x 6″	1.000	S.F.	.028	1.49	.81	2.30
1/2″ x 8″	1.000	S.F.	.023	1.49	.67	2.16
Redwood, clear vertical grain, 1/2″ x 6″	1.000	S.F.	.028	1.98	.81	2.79
1/2″ x 8″	1.000	S.F.	.032	2.11	.94	3.05
Siding board & batten, cedar, "B" grade, 1″ x 10″	1.000	S.F.	.031	2.09	.90	2.99
1″ x 12″	1.000	S.F.	.031	2.09	.90	2.99
White pine, #2 & better, 1″ x 10″	1.000	S.F.	.029	.73	.85	1.58
1″ x 12″	1.000	S.F.	.029	.73	.85	1.58
Siding vertical, tongue & groove, cedar "B" grade, 1″ x 4″	1.000	S.F.	.033	1.21	.98	2.19
1″ x 8″	1.000	S.F.	.024	1.21	.69	1.90
"A" grade, 1″ x 4″	1.000	S.F.	.033	1.60	.98	2.58
1″ x 8″	1.000	S.F.	.024	1.60	.69	2.29
Clear vertical grain, 1″ x 4″	1.000	S.F.	.033	2.37	.98	3.35
1″ x 6″	1.000	S.F.	.024	2.37	.69	3.06
1″ x 8″	1.000	S.F.	.024	2.31	.69	3.00
Siding plywood, texture 1-11 cedar, 3/8″ thick	1.000	S.F.	.024	1.19	.64	1.83
5/8″ thick	1.000	S.F.	.024	1.56	.64	2.20
Redwood, 3/8″ thick	1.000	S.F.	.024	1.19	.64	1.83
5/8″ thick	1.000	S.F.	.024	1.86	.64	2.50
Fir, 3/8″ thick	1.000	S.F.	.024	.62	.64	1.26
5/8″ thick	1.000	S.F.	.024	1.06	.64	1.70
Hard board, 7/16″ thick primed, plain finish	1.000	S.F.	.025	.78	.66	1.44

Shingle Siding Systems

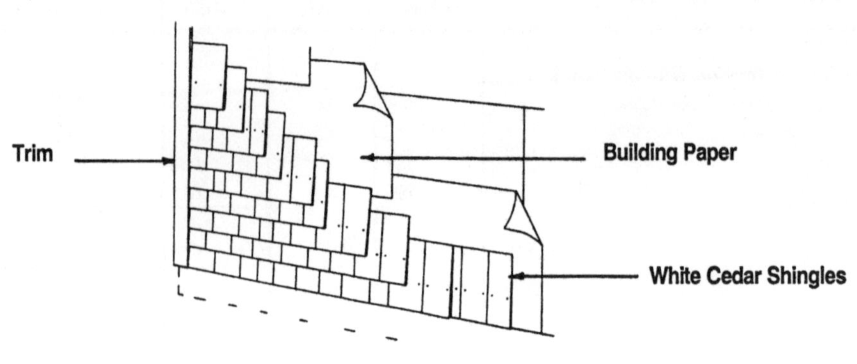

System Description	QUAN.	UNIT	LABOR-HOURS	COST PER S.F.			S.F. EXTENSIONS
				MAT.	INST.	TOTAL	
WHITE CEDAR SHINGLES, 5″ EXPOSURE							
White cedar shingles, 16″ long, grade "A", 5″ exposure	1.000	S.F.	.033	.99	.98	1.97	
#15 asphalt felt paper	1.100	S.F.	.002	.03	.07	.10	
Trim, cedar	.125	S.F.	.005	.17	.15	.32	
Paint, primer & 1 coat	1.000	S.F.	.017	.07	.46	.53	
TOTAL			.057	1.26	1.66	2.92	
WHITE CEDAR SHINGLES, 5″ EXPOSURE, GRADE "B"							
White cedar shingles, 16″ long, grade "B", 5″ exp.	1.000	S.F.	.040	.94	1.17	2.11	
#15 asphalt felt paper	1.100	S.F.	.002	.03	.07	.10	
Trim, cedar	.125	S.F.	.005	.17	.15	.32	
Paint, primer & 1 coat	1.000	S.F.	.017	.07	.46	.53	
TOTAL			.064	1.21	1.85	3.06	
FIRE RETARDANT, WHITE CEDAR SHINGLES, 5″ EXPOSURE							
White cedar shingles, 16″ long, grade "A", 5″ exp.	1.000	S.F.	.033	1.39	.98	2.37	
#15 asphalt felt paper	1.100	S.F.	.002	.03	.07	.10	
Trim, cedar	.125	S.F.	.005	.17	.15	.32	
Paint, primer & 1 coat	1.000	S.F.	.017	.07	.46	.53	
TOTAL			.057	1.66	1.66	3.32	
NO. 1 PERFECTIONS, 5-1/2″ EXPOSURE							
No. 1 perfections, red cedar, 5-1/2″ exposure	1.000	S.F.	.029	1.98	.86	2.84	
#15 asphalt felt paper	1.100	S.F.	.002	.03	.07	.10	
Trim, cedar	.125	S.F.	.005	.17	.15	.32	
Stain, sealer & 1 coat	1.000	S.F.	.017	.07	.46	.53	
TOTAL			.053	2.25	1.54	3.79	
NO. 1 PERFECTIONS, 10″ EXPOSURE							
No. 1 perfections, red cedar, 10″ exposure	1.000	S.F.	.025	1.02	.73	1.75	
#15 asphalt felt paper	1.100	S.F.	.002	.03	.07	.10	
Trim, cedar	.125	S.F.	.005	.17	.15	.32	
Stain, sealer & 1 coat	1.000	S.F.	.017	.07	.46	.53	
TOTAL			.049	1.29	1.41	2.70	

Shingle Siding Systems

System Description	QUAN.	UNIT	LABOR-HOURS	COST PER S.F.			S.F. EXTENSIONS
				MAT.	INST.	TOTAL	
RESQUARED & REBUTTED PERFECTIONS, 5-1/2″ EXPOSURE							
Resquared & rebutted perfections, 5-1/2″ exposure	1.000	S.F.	.027	1.10	.79	1.89	
#15 asphalt felt paper	1.100	S.F.	.002	.03	.07	.10	
Trim, cedar	.125	S.F.	.005	.17	.15	.32	
Stain, sealer & 1 coat	1.000	S.F.	.017	.07	.46	.53	
TOTAL			.051	1.37	1.47	2.84	
HAND-SPLIT SHAKES, 8-1/2″ EXPOSURE							
Hand-split red cedar shakes, 18″ long, 8-1/2″ exposure	1.000	S.F.	.040	1.51	1.17	2.68	
#15 asphalt felt paper	1.100	S.F.	.002	.03	.07	.10	
Trim, cedar	.125	S.F.	.005	.17	.15	.32	
Stain, sealer & 1 coat	1.000	S.F.	.017	.07	.46	.53	
TOTAL			.064	1.78	1.85	3.63	
FIRE RETARDANT, WOOD SHAKES, 8-1/2″ EXPOSURE							
Hand-split red cedar shakes, 18″ long, 8-1/2″ exp.	1.000	S.F.	.038	2.48	1.13	3.61	
#15 asphalt felt paper	1.100	S.F.	.002	.03	.07	.10	
Trim, cedar	.125	S.F.	.005	.17	.15	.32	
Stain, sealer & 1 coat	1.000	S.F.	.017	.07	.46	.53	
TOTAL			.062	2.75	1.81	4.56	

The costs in this system are based on a square foot of wall area.

Shingle Siding System Components

Component Description	QUAN.	UNIT	LABOR-HOURS	COST PER S.F.		
				MAT.	INST.	TOTAL
Shingles						
Shingles wood, white cedar 16″ long, "A" grade, 5″ exposure	1.000	S.F.	.033	.99	.98	1.97
7″ exposure	1.000	S.F.	.030	.89	.88	1.77
8-1/2″ exposure	1.000	S.F.	.032	.56	.94	1.50
10″ exposure	1.000	S.F.	.028	.49	.82	1.31
"B" grade, 5″ exposure	1.000	S.F.	.040	.94	1.17	2.11
7″ exposure	1.000	S.F.	.028	.65	.82	1.47
8-1/2″ exposure	1.000	S.F.	.024	.56	.70	1.26
10″ exposure	1.000	S.F.	.020	.47	.59	1.06
Fire retardant, "A" grade, 5″ exposure	1.000	S.F.	.033	1.39	.98	2.37
7″ exposure	1.000	S.F.	.030	1.25	.88	2.13
8-1/2″ exposure	1.000	S.F.	.027	1.11	.78	1.89
10″ exposure	1.000	S.F.	.023	.97	.69	1.66
"B" grade, 5″ exposure	1.000	S.F.	.040	1.34	1.17	2.51
7″ exposure	1.000	S.F.	.028	.93	.82	1.75
8-1/2″ exposure	1.000	S.F.	.024	.80	.70	1.50
10″ exposure	1.000	S.F.	.020	.67	.59	1.26
Hand-split, red cedar, 24″ long, 7″ exposure	1.000	S.F.	.045	2.31	1.32	3.63
8-1/2″ exposure	1.000	S.F.	.038	1.98	1.13	3.11
10″ exposure	1.000	S.F.	.032	1.65	.94	2.59
12″ exposure	1.000	S.F.	.026	1.32	.75	2.07
Fire retardant, 7″ exposure	1.000	S.F.	.090	5.20	2.64	7.84
18″ long, 5″ exposure	1.000	S.F.	.068	2.57	1.99	4.56
7″ exposure	1.000	S.F.	.048	1.81	1.40	3.21
10″ exposure	1.000	S.F.	.036	1.36	1.05	2.41

Metal & Plastic Siding Systems

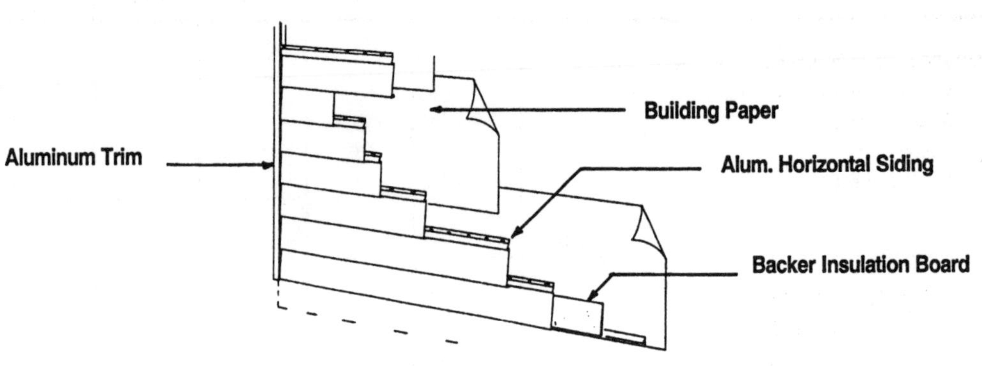

System Description	QUAN.	UNIT	LABOR-HOURS	COST PER S.F.			S.F. EXTENSIONS
				MAT.	INST.	TOTAL	
ALUMINUM CLAPBOARD SIDING, 8" WIDE, WHITE							
Aluminum horizontal siding, 8" clapboard	1.000	S.F.	.031	1.29	.84	2.13	
Backer, insulation board	1.000	S.F.	.008	.32	.23	.55	
Trim, aluminum	.600	L.F.	.016	.64	.42	1.06	
Paper, #15 asphalt felt	1.100	S.F.	.002	.03	.07	.10	
TOTAL			.057	2.28	1.56	3.84	
ALUMINUM CLAPBOARD SIDING, DOUBLE 5" PATTERN							
Aluminum horizontal siding, 10" clapboard, white	1.000	S.F.	.029	1.22	.79	2.01	
Backer, insulation board	1.000	S.F.	.008	.32	.23	.55	
Trim, Aluminum	.600	L.F.	.016	.64	.42	1.06	
Paper, #15 asphalt felt	1.100	S.F.	.002	.03	.07	.10	
TOTAL			.055	2.21	1.51	3.72	
ALUMINUM CLAPBOARD SIDING, SHAKE FINISH							
Aluminum horizontal siding, 10" shake finish, white	1.000	S.F.	.029	1.53	.79	2.32	
Backer, insulation board	1.000	S.F.	.008	.32	.23	.55	
Trim, Aluminum	.600	L.F.	.016	.64	.42	1.06	
Paper, #15 asphalt felt	1.100	S.F.	.002	.03	.07	.10	
TOTAL			.055	2.52	1.51	4.03	
ALUMINUM VERTICAL BOARD & BATTEN, WHITE							
Aluminum vertical board & batten	1.000	S.F.	.027	1.46	.73	2.19	
Backer insulation board	1.000	S.F.	.008	.32	.23	.55	
Trim, aluminum	.600	L.F.	.016	.64	.42	1.06	
Paper, #15 asphalt felt	1.100	S.F.	.002	.03	.07	.10	
TOTAL			.053	2.45	1.45	3.90	
VINYL CLAPBOARD SIDING, 8" WIDE, WHITE							
PVC vinyl horizontal siding, 8" clapboard	1.000	S.F.	.032	.75	.87	1.62	
Backer, insulation board	1.000	S.F.	.008	.32	.23	.55	
Trim, vinyl	.600	L.F.	.014	.38	.37	.75	
Paper, #15 asphalt felt	1.100	S.F.	.002	.03	.07	.10	
TOTAL			.056	1.48	1.54	3.02	

Metal & Plastic Siding Systems

System Description	QUAN.	UNIT	LABOR-HOURS	COST PER S.F.			S.F. EXTENSIONS
				MAT.	INST.	TOTAL	
VINYL CLAPBOARD SIDING, DOUBLE 5" PATTERN							
PVC vinyl horizontal siding, 10" clapboard, white	1.000	S.F.	.029	.78	.79	1.57	
Backer, insulation board	1.000	S.F.	.008	.32	.23	.55	
Trim, vinyl	.600	L.F.	.014	.38	.37	.75	
Paper, #15 asphalt felt	1.100	S.F.	.002	.03	.07	.10	
TOTAL			.053	1.51	1.46	2.97	
VINYL CLAPBOARD SIDING, EMBOSSED, SINGLE							
PVC vinyl horizontal siding, embossed 8", white	1.000	S.F.	.032	.76	.87	1.63	
Backer, insulation board	1.000	S.F.	.008	.32	.23	.55	
Trim, vinyl	.600	L.F.	.014	.38	.37	.75	
Paper, #15 asphalt felt	1.100	S.F.	.002	.03	.07	.10	
TOTAL			.056	1.49	1.54	3.03	
VINYL VERTICAL BOARD & BATTEN, WHITE							
PVC vinyl vertical board & batten	1.000	S.F.	.029	1.23	.79	2.02	
Backer, insulation board	1.000	S.F.	.008	.32	.23	.55	
Trim, vinyl	.600	L.F.	.014	.38	.37	.75	
Paper, #15 asphalt felt	1.100	S.F.	.002	.03	.07	.10	
TOTAL			.053	1.96	1.46	3.42	

The costs in this system are on a square foot of wall basis.

Metal & Plastic Siding System Components

Component Description	QUAN.	UNIT	LABOR-HOURS	COST PER S.F.		
				MAT.	INST.	TOTAL
Siding						
Siding, aluminum, .024" thick, smooth, 8" wide, white	1.000	S.F.	.031	1.29	.84	2.13
Color	1.000	S.F.	.031	1.37	.84	2.21
Double 4" pattern, 8" wide, white	1.000	S.F.	.031	1.22	.84	2.06
Color	1.000	S.F.	.031	1.30	.84	2.14
Embossed, single, 8" wide, white	1.000	S.F.	.031	1.45	.84	2.29
Color	1.000	S.F.	.031	1.53	.84	2.37
Alum siding with insulation board, smooth, 8" wide, white	1.000	S.F.	.031	1.24	.84	2.08
Color	1.000	S.F.	.031	1.32	.84	2.16
Vinyl siding, 8" wide, smooth, white	1.000	S.F.	.032	.75	.87	1.62
Color	1.000	S.F.	.032	.82	.87	1.69
10" wide, smooth, white	1.000	S.F.	.029	.86	.79	1.65
Color	1.000	S.F.	.029	.93	.79	1.72
Vinyl, shake finish, 10" wide, white	1.000	S.F.	.029	1.67	.79	2.46
Color	1.000	S.F.	.029	1.74	.79	2.53
Backer board						
Backer board, installed in siding panels 8" or 10" wide	1.000	S.F.	.008	.32	.23	.55
4' x 8' sheets, polystyrene, 3/4" thick	1.000	S.F.	.010	.34	.29	.63
4' x 8' fiberboard, plain	1.000	S.F.	.008	.32	.23	.55
Foil faced	1.000	S.F.	.012	.28	.35	.63
Paper						
Kraft paper, plain	1.100	S.F.	.002	.04	.07	.11
Foil backed	1.100	S.F.	.002	.08	.07	.15

Insulation Systems

COMPONENT DESCRIPTION	QUAN.	UNIT	LABOR-HOURS	COST PER S.F.		
				MAT.	INST.	TOTAL
Poured Insulation						
Poured insulation, cellulose fiber, R3.8 per inch (1'' thick)	1.000	S.F.	.003	.04	.10	.14
Fiberglass , R4.0 per inch (1'' thick)	1.000	S.F.	.003	.03	.10	.13
Mineral wool, R3.0 per inch (1'' thick)	1.000	S.F.	.003	.03	.10	.13
Polystyrene, R4.0 per inch (1'' thick)	1.000	S.F.	.003	.16	.10	.26
Vermiculite, R2.7 per inch (1'' thick)	1.000	S.F.	.003	.14	.10	.24
Perlite, R2.7 per inch (1'' thick)	1.000	S.F.	.003	.14	.10	.24
Reflective Insulation						
Reflective, aluminum foil on kraft paper, foil one side R9	1.000	S.F.	.004	.04	.12	.16
Multilayered with air spaces, 2 ply, R14	1.000	S.F.	.004	.17	.12	.29
3 ply, R17	1.000	S.F.	.005	.22	.16	.38
5 ply, R22	1.000	S.F.	.005	.32	.16	.48
Rigid Insulation						
Rigid insulation, fiberglass, unfaced,						
1-1/2'' thick, R6.2	1.000	S.F.	.008	.32	.23	.55
2'' thick, R8.3	1.000	S.F.	.008	.37	.23	.60
2-1/2'' thick, R10.3	1.000	S.F.	.010	.50	.29	.79
3'' thick, R12.4	1.000	S.F.	.010	.50	.29	.79
Foil faced, 1'' thick, R4.3	1.000	S.F.	.008	.76	.23	.99
1-1/2'' thick, R6.2	1.000	S.F.	.008	1.02	.23	1.25
2'' thick, R8.7	1.000	S.F.	.009	1.29	.26	1.55
2-1/2'' thick, R10.9	1.000	S.F.	.010	1.52	.29	1.81
3'' thick, R13.0	1.000	S.F.	.010	1.65	.29	1.94
Foam glass, 1-1/2'' thick R2.64	1.000	S.F.	.010	1.41	.29	1.70
2'' thick R5.26	1.000	S.F.	.011	2.80	.32	3.12
Perlite, 1'' thick R2.77	1.000	S.F.	.010	.34	.29	.63
2'' thick R5.55	1.000	S.F.	.011	.63	.32	.95
Polystyrene, extruded, blue, 2.2#/C.F., 3/4'' thick R4	1.000	S.F.	.010	.34	.29	.63
1-1/2'' thick R8.1	1.000	S.F.	.011	.63	.32	.95
2'' thick R10.8	1.000	S.F.	.011	.91	.32	1.23
Molded bead board, white, 1'' thick R3.85	1.000	S.F.	.010	.20	.29	.49
1-1/2'' thick, R5.6	1.000	S.F.	.011	.40	.32	.72
2'' thick, R7.7	1.000	S.F.	.011	.59	.32	.91
Non-rigid insulation, batts						
Fiberglass, kraft faced, 3-1/2'' thick, R11, 11'' wide	1.000	S.F.	.005	.22	.15	.37
15'' wide	1.000	S.F.	.005	.22	.15	.37
23'' wide	1.000	S.F.	.005	.22	.15	.37
6'' thick, R19, 11'' wide	1.000	S.F.	.006	.29	.17	.46
15'' wide	1.000	S.F.	.006	.29	.17	.46
23'' wide	1.000	S.F.	.006	.29	.17	.46
9'' thick, R30, 15'' wide	1.000	S.F.	.006	.52	.17	.69
23'' wide	1.000	S.F.	.006	.52	.17	.69
12'' thick, R38, 15'' wide	1.000	S.F.	.006	.71	.17	.88
23'' wide	1.000	S.F.	.006	.71	.17	.88
Fiberglass, foil faced, 3-1/2'' thick, R11, 15'' wide	1.000	S.F.	.005	.26	.15	.41
23'' wide	1.000	S.F.	.005	.26	.15	.41
6'' thick, R19, 15'' thick	1.000	S.F.	.005	.33	.15	.48
23'' wide	1.000	S.F.	.005	.33	.15	.48
9'' thick, R30, 15'' wide	1.000	S.F.	.006	.59	.17	.76
Non-rigid insulation batts						
Fiberglass unfaced, 3-1/2'' thick, R11, 15'' wide	1.000	S.F.	.005	.20	.15	.35
23'' wide	1.000	S.F.	.005	.20	.15	.35
6'' thick, R19, 15'' wide	1.000	S.F.	.006	.33	.17	.50

Insulation Systems

Component Descriptions	QUAN.	UNIT	LABOR-HOURS	COST PER S.F.		
				MAT.	INST.	TOTAL
23″ wide	1.000	S.F.	.006	.33	.17	.50
9″ thick, R19, 15″ wide	1.000	S.F.	.007	.55	.20	.75
23″ wide	1.000	S.F.	.007	.55	.20	.75
12″ thick, R38, 15″ wide	1.000	S.F.	.007	.70	.20	.90
23″ wide	1.000	S.F.	.007	.70	.20	.90
Mineral fiber batts, 3″ thick, R11	1.000	S.F.	.005	.35	.15	.50
3-1/2″ thick, R13	1.000	S.F.	.005	.35	.15	.50
6″ thick, R19	1.000	S.F.	.005	.51	.15	.66
6-1/2″ thick, R22	1.000	S.F.	.005	.51	.15	.66
10″ thick, R30	1.000	S.F.	.006	.86	.17	1.03

Double Hung Window Systems

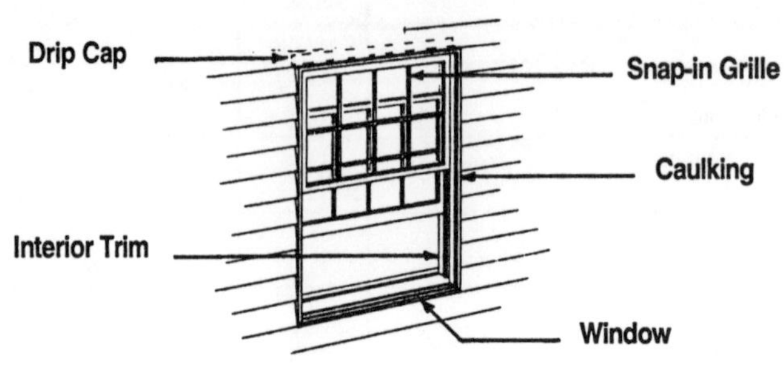

System Description	QUAN.	UNIT	LABOR-HOURS	MAT.	INST.	TOTAL	EA. EXTENSIONS
BUILDER'S QUALITY WOOD WINDOW 2' X 3', DOUBLE HUNG							
Window, primed, builder's quality, 2' x 3', insulating glass	1.000	Ea.	.800	140.00	23.50	163.50	
Trim, interior casing	11.000	L.F.	.367	8.69	10.78	19.47	
Paint, interior & exterior, primer & 2 coats	2.000	Face	1.600	1.74	42.00	43.74	
Caulking	10.000	L.F.	.308	.90	9.10	10.00	
Snap-in grille	1.000	Set	.333	19.30	9.80	29.10	
Drip cap, metal	2.000	L.F.	.040	.42	1.18	1.60	
TOTAL			3.448	171.05	96.36	267.41	
BUILDER'S QUALITY WOOD WINDOW 3' X 4' DOUBLE HUNG							
Window, builder's quality, 3' x 4', insulating glass	1.000	Ea.	.889	181.00	26.00	207.00	
Trim, interior casing	15.000	L.F.	.367	8.69	10.78	19.47	
Paint, interior & exterior, primer & 2 coats	2.000	Face	1.600	1.74	42.00	43.74	
Caulking	14.000	L.F.	.308	.90	9.10	10.00	
Snap-in grille	1.000	Set	.333	19.30	9.80	29.10	
Drip cap, metal	3.000	L.F.	.040	.42	1.18	1.60	
TOTAL			3.537	212.05	98.86	310.91	
PLASTIC CLAD WOOD WINDOW 3' X 4', DOUBLE HUNG							
Window, plastic clad, premium, 3' x 4', insulating glass	1.000	Ea.	.889	330.00	26.00	356.00	
Trim, interior casing	15.000	L.F.	.500	11.85	14.70	26.55	
Paint, interior, primer & 2 coats	1.000	Face	.800	.87	21.00	21.87	
Caulking	14.000	L.F.	.431	1.26	12.74	14.00	
Snap-in grille	1.000	Set	.333	19.30	9.80	29.10	
TOTAL			2.953	363.28	84.24	447.52	
METAL CLAD WOOD WINDOW 3' X 4' DOUBLE HUNG							
Window, deluxe, 3' x 4' insulating glass	1.000	Ea.	.889	200.00	26.00	226.00	
Trim interior casing	15.000	L.F.	.567	13.43	16.66	30.09	
Paint interior, primer & 2 coats	1.000	Face	.800	.87	21.00	21.87	
Caulking	14.000	L.F.	.492	1.44	14.56	16.00	
Snap-in grille	1.000	Set	.235	44.50	6.90	51.40	
Drip cap, metal	3.000	L.F.	.060	.63	1.77	2.40	
TOTAL			3.043	260.87	86.89	347.76	

Double Hung Window Systems

System Description	QUAN.	UNIT	LABOR-HOURS	COST EACH			EA. EXTENSIONS
				MAT.	INST.	TOTAL	
PLASTIC CLAD WOOD WINDOW 3'-6'' X 6' DOUBLE HUNG							
Window, premium 3'-6'' x 6' insulating glass	1.000	Ea.	1.000	460.00	29.50	489.50	
Trim interior casing	20.000	L.F.	.667	15.80	19.60	35.40	
Paint, interior, primer & 2 coats	1.000	Face	.800	.87	21.00	21.87	
Caulking	19.000	L.F.	.585	1.71	17.29	19.00	
Snap-in grille	1.000	Set	.235	44.50	6.90	51.40	
TOTAL			3.287	522.88	94.29	617.17	
METAL CLAD WOOD WINDOW, 3' X 5', DOUBLE HUNG							
Window, metal clad, deluxe, 3' x 5', insulating glass	1.000	Ea.	1.000	226.00	29.50	255.50	
Trim, interior casing	17.000	L.F.	.567	13.43	16.66	30.09	
Paint, interior, primer & 2 coats	1.000	Face	.800	.87	21.00	21.87	
Caulking	16.000	L.F.	.492	1.44	14.56	16.00	
Snap-in grille	1.000	Set	.235	44.50	6.90	51.40	
Drip cap, metal	3.000	L.F.	.060	.63	1.77	2.40	
TOTAL			3.154	286.87	90.39	377.26	
BUILDER'S QUALITY WOOD WINDOW 4' X 4'-6'' DOUBLE HUNG							
Window, builder's quality, 4' x 4'-6'', insulating glass	1.000	Ea.	1.000	215.00	29.50	244.50	
Trim, interior casing	18.000	L.F.	.600	14.22	17.64	31.86	
Paint interior & exterior, primer & 2 coats	2.000	Face	1.600	1.74	42.00	43.74	
Caulking	17.000	L.F.	.523	1.53	15.47	17.00	
Snap-in grille	1.000	Set	.235	44.50	6.90	51.40	
Drip cap, metal	4.000	L.F.	.080	.84	2.36	3.20	
TOTAL			4.038	277.83	113.87	391.70	
PLASTIC CLAD WOOD WINDOW 2'-6'' X 3' DOUBLE HUNG							
Window, premium, 2'-6'' x 3', insulating glass	1.000	Ea.	.800	140.00	23.50	163.50	
Trim interior casing	12.000	m&L.F.	.400	9.48	11.76	21.24	
Paint, interior, primer & 2 coats	1.000	Face	.800	.87	21.00	21.87	
Caulking	11.000	L.F.	.338	.99	10.01	11.00	
Snap-in grille	1.000	Set	.333	19.30	9.80	29.10	
TOTAL			2.671	170.64	76.07	246.71	

The cost of this system is on a cost per each window basis.

Double Hung Window System Components

Component Description	QUAN.	UNIT	LABOR-HOURS	COST EACH		
				MAT.	INST.	TOTAL
Windows, double-hung						
Windows, double-hung, builder's quality, 2' x 3', single glass	1.000	Ea.	.800	99.50	23.50	123.00
3' x 4', single glass	1.000	Ea.	.889	132.00	26.00	158.00
Plastic clad premium insulating glass, 2'-6'' x 3'	1.000	Ea.	.800	240.00	23.50	263.50
3' x 3'-6''	1.000	Ea.	.800	305.00	23.50	328.50
Metal clad deluxe insulating glass, 2'-6'' x 3'	1.000	Ea.	.800	157.00	23.50	180.50
3' x 3'-6''	1.000	Ea.	.800	187.00	23.50	210.50
Trim, interior casing						
Trim, interior casing, window 2' x 3'	11.000	L.F.	.367	8.70	10.80	19.50
3' x 3'-6''	14.000	L.F.	.467	11.05	13.70	24.75

Casement Window Systems

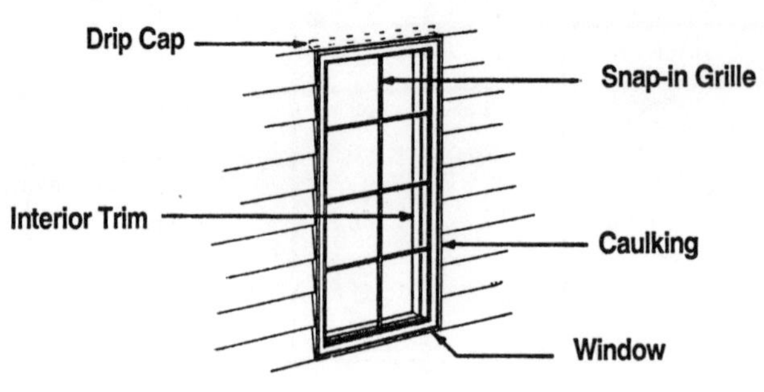

System Description	QUAN.	UNIT	LABOR-HOURS	COST EACH			EA. EXTENSIONS
				MAT.	INST.	TOTAL	
BUILDER'S QUALITY WINDOW, WOOD, 2' BY 3', CASEMENT							
Window, primed, builder's quality, 2' x 3', insulating glass	1.000	Ea.	.800	204.00	23.50	227.50	
Trim, interior casing	11.000	L.F.	.367	8.69	10.78	19.47	
Paint, interior & exterior, primer & 2 coats	2.000	Face	1.600	1.74	42.00	43.74	
Caulking	10.000	L.F.	.308	.90	9.10	10.00	
Snap-in grille	1.000	Ea.	.267	13.90	7.85	21.75	
Drip cap, metal	2.000	L.F.	.040	.42	1.18	1.60	
TOTAL			3.382	229.65	94.41	324.06	
PLASTIC CLAD WOOD WINDOW 2' X 3' CASEMENT							
Window, premium, 2' x 3', insulating glass	1.000	Ea.	.800	288.00	23.50	311.50	
Trim, interior casing	11.000	L.F.	.367	8.69	10.78	19.47	
Paint, interior, primer & 2 coats	1.000	Face	.800	.87	21.00	21.87	
Caulking	10.000	L.F.	.308	.90	9.10	10.00	
Snap-in grille	1.000	Set	.267	13.90	7.85	21.75	
TOTAL			2.542	312.36	72.23	384.59	
METAL CLAD WOOD WINDOW 2' X 3' CASEMENT							
Window, deluxe, 2' x 3' insulating glass	1.000	Ea.	.800	525.00	23.50	548.50	
Trim, interior casing	11.000	L.F.	.367	8.69	10.78	19.47	
Paint, interior, primer & 2 coats	1.000	Face	.800	.87	21.00	21.87	
Caulking	10.000	L.F.	.308	.90	9.10	10.00	
Snap-in grille	1.000	Set	.267	13.90	7.85	21.75	
Drip cap, metal	2.000	L.F.	.040	.42	1.18	1.60	
TOTAL			2.582	549.78	73.41	623.19	
BUILDER'S QUALITY WOOD WINDOW 2' X 4'-6'' CASEMENT							
Window, builder's quality, 2' x 4'-6", insulating glass	1.000	Ea.	.727	495.00	21.50	516.50	
Trim, interior casing	14.000	L.F.	.467	11.06	13.72	24.78	
Paint interior & exterior, primer & 2 coats	2.000	Face	1.600	1.74	42.00	43.74	
Caulking	13.000	L.F.	.400	1.17	11.83	13.00	
Snap-in grille	1.000	Set	.250	28.50	7.35	35.85	
Drip cap, metal	2.000	L.F.	.040	.42	1.18	1.60	
TOTAL			3.484	537.89	97.58	635.47	

Casement Window Systems

System Description	QUAN.	UNIT	LABOR-HOURS	COST EACH			EA. EXTENSIONS
				MAT.	INST.	TOTAL	
PLASTIC CLAD WOOD WINDOW, 2' X 4', CASEMENT							
Window, plastic clad, premium, 2' x 4', insulating glass	1.000	Ea.	.889	220.00	26.00	246.00	
Trim, interior casing	13.000	L.F.	.433	10.27	12.74	23.01	
Paint, interior, primer & 2 coats	1.000	Ea.	.800	.87	21.00	21.87	
Caulking	12.000	L.F.	.369	1.08	10.92	12.00	
Snap-in grille	1.000	Ea.	.267	13.90	7.85	21.75	
TOTAL			2.758	246.12	78.51	324.63	
METAL CLAD WOOD WINDOW 2' X 4' CASEMENT							
Window, deluxe, 2' x 4', insulating glass	1.000	Ea.	.889	281.00	26.00	307.00	
Trim, interior casing	13.000	L.F.	.433	10.27	12.74	23.01	
Paint, interior, primer & 2 coats	1.000	Face	.800	.87	21.00	21.87	
Caulking	12.000	L.F.	.369	1.08	10.92	12.00	
Snap-in grille	1.000	Set	.267	13.90	7.85	21.75	
Drip cap, metal	2.000	L.F.	.040	.42	1.18	1.60	
TOTAL			2.798	307.54	79.69	387.23	
BUILDER'S QUALITY WOOD WINDOW 2' X 6' CASEMENT							
Window, builder's quality, 2' x 6' insulating glass	1.000	Ea.	.889	810.00	26.00	836.00	
Trim, interior casing	17.000	L.F.	.567	13.43	16.66	30.09	
Paint, interior & exterior, primer & 2 coats	2.000	Face	1.600	1.74	42.00	43.74	
Caulking	16.000	L.F.	.492	1.44	14.56	16.00	
Snap-in grille	1.000	Set	.250	28.50	7.35	35.85	
Drip cap, metal	2.000	L.F.	.040	.42	1.18	1.60	
TOTAL			3.838	855.53	107.75	963.28	
METAL CLAD WOOD WINDOW 2' X 6' CASEMENT							
Window, deluxe, 2' x 6', insulating glass	1.000	Ea.	1.000	330.00	29.50	359.50	
Trim, interior casing	17.000	L.F.	.567	13.43	16.66	30.09	
Paint, interior, primer & 2 coats	1.000	Face	.800	.87	21.00	21.87	
Caulking	16.000	L.F.	.492	1.44	14.56	16.00	
Snap-in grille	1.000	Set	.250	28.50	7.35	35.85	
Drip cap, metal	2.000	L.F.	.040	.42	1.18	1.60	
TOTAL			3.149	374.66	90.25	464.91	

The cost of this system is on a cost per each window basis.

Casement Window System Components

Component Description	QUAN.	UNIT	LABOR-HOURS	COST EACH		
				MAT.	INST.	TOTAL
Window, casement						
Window, casement, builders quality, 2' x 3', single glass	1.000	Ea.	.800	193.00	23.50	216.50
2' x 4'-6'', single glass	1.000	Ea.	.727	495.00	21.50	516.50
Plastic clad premium insulating glass, 2' x 3'	1.000	Ea.	.800	288.00	23.50	311.50
2' x 5'	1.000	Ea.	1.000	264.00	29.50	293.50
Paint or stain						
Paint or stain, interior or exterior, 2' x 3' window, 1 coat	1.000	Face	.381	.32	10.00	10.32
2 coats	1.000	Face	.615	.63	16.20	16.83

Awning Window Systems

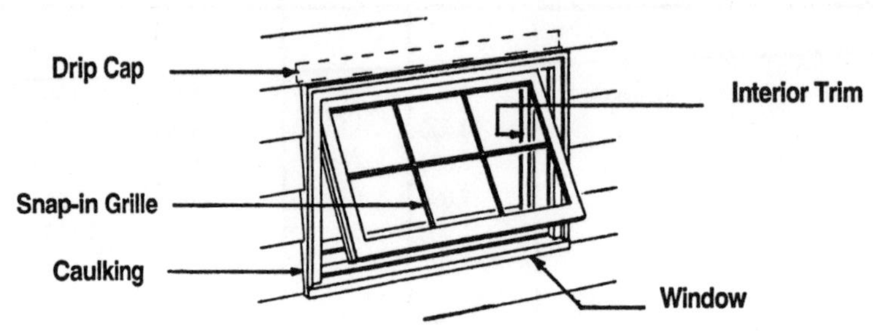

System Description	QUAN.	UNIT	LABOR-HOURS	COST EACH			EA. EXTENSIONS
				MAT.	INST.	TOTAL	
BUILDER'S QUALITY WINDOW, WOOD, 34″ X 22″, AWNING							
Window, builder quality, 34″ x 22″, insulating glass	1.000	Ea.	.800	223.00	23.50	246.50	
Trim, interior casing	10.500	L.F.	.350	8.30	10.29	18.59	
Paint, interior & exterior, primer & 2 coats	2.000	Face	1.600	1.74	42.00	43.74	
Caulking	9.500	L.F.	.292	.86	8.65	9.51	
Snap-in grille	1.000	Ea.	.267	18.40	7.85	26.25	
Drip cap, metal	3.000	L.F.	.060	.63	1.77	2.40	
TOTAL			3.369	252.93	94.06	346.99	
PLASTIC CLAD WOOD WINDOW 34″ X 22″ AWNING							
Window, premium 34″ x 22″, insulating glass	1.000	Ea.	.800	193.00	23.50	216.50	
Trim, interior casing	10.500	L.F.	.350	8.30	10.29	18.59	
Paint, interior, primer & 2 coats	1.000	Face	.800	.87	21.00	21.87	
Caulking	9.500	L.F.	.292	.86	8.65	9.51	
Snap-in grille	1.000	Set	.267	18.40	7.85	26.25	
TOTAL			2.509	221.43	71.29	292.72	
METAL CLAD WOOD WINDOW 34″ X 22″ AWNING							
Window, deluxe, 34″ x 22″, insulating glass	1.000	Ea.	.800	175.00	23.50	198.50	
Trim, interior casing	10.500	L.F.	.350	8.30	10.29	18.59	
Paint, interior, primer & 2 coats	1.000	Face	.800	.87	21.00	21.87	
Caulking	9.500	L.F.	.292	.86	8.65	9.51	
Snap-in grille	1.000	Ea.	.267	18.40	7.85	26.25	
Drip cap, metal	3.000		.060	.63	1.77	2.40	
TOTAL			2.569	204.06	73.06	277.12	
BUILDER'S QUALITY WINDOW, WOOD, 40″ X 28″, AWNING							
Window, builder's quality, 40″ x 28″, insulating glass	1.000	Ea.	.889	270.00	26.00	296.00	
Trim, interior casing	13.500	L.F.	.450	10.67	13.23	23.90	
Paint interior & exterior primer & 2 coats	2.000	Face	1.600	1.74	42.00	43.74	
Caulking	12.500	L.F.	.385	1.13	11.38	12.51	
Snap-in grille	1.000	Set	.250	18.95	7.35	26.30	
Drip cap, metal	3.500	L.F.	.080	.84	2.36	3.20	
TOTAL			3.654	303.33	102.32	405.65	

Awning Window Systems

System Description	QUAN.	UNIT	LABOR-HOURS	COST EACH			EA. EXTENSIONS
				MAT.	INST.	TOTAL	
PLASTIC CLAD WOOD WINDOW, 40" X 28", AWNING							
Window, plastic clad, premium, 40" x 28", insulating glass	1.000	Ea.	.889	310.00	26.00	336.00	
Trim interior casing	13.500	L.F.	.450	10.67	13.23	23.90	
Paint, interior, primer & 2 coats	1.000	Face	.800	.87	21.00	21.87	
Caulking	12.500	L.F.	.385	1.13	11.38	12.51	
Snap-in grille	1.000	Ea.	.267	18.40	7.85	26.25	
TOTAL			2.791	341.07	79.46	420.53	
METAL CLAD WOOD WINDOW 40" X 28" AWNING							
Window, deluxe, 40" x 28", insulating glass	1.000	Ea.	.889	209.00	26.00	235.00	
Trim, interior casing	13.500	L.F.	.450	10.67	13.23	23.90	
Paint, interior, primer & 2 coats	1.000	Face	.800	.87	21.00	21.87	
Caulking	12.500	L.F.	.385	1.13	11.38	12.51	
Snap-in grille	1.000	Set	.267	18.40	7.85	26.25	
Drip cap, metal	3.500	L.F.	.080	.84	2.36	3.20	
TOTAL			2.871	240.91	81.82	322.73	
PLASTIC CLAD WOOD WINDOW 48" X 36" AWNING							
Window, premium, 48" x 36" insulating glass	1.000	Ea.	1.000	269.00	29.50	298.50	
Trim, interior casing	15.000	L.F.	.500	11.85	14.70	26.55	
Paint, interior, primer & 2 coats	1.000	Face	.800	.87	21.00	21.87	
Caulking	14.000	L.F.	.431	1.26	12.74	14.00	
Snap-in grille	1.000	Ea.	.250	18.95	7.35	26.30	
TOTAL			2.981	301.93	85.29	387.22	
METAL CLAD WOOD WINDOW, 48" X 36", AWNING							
Window, metal clad, deluxe, 48" x 36", insulating glass	1.000	Ea.	1.000	253.00	29.50	282.50	
Trim, interior casing	15.000	L.F.	.500	11.85	14.70	26.55	
Paint, interior, primer & 2 coats	1.000	Face	.800	.87	21.00	21.87	
Caulking	14.000	L.F.	.431	1.26	12.74	14.00	
Snap-in grille	1.000	Ea.	.250	18.95	7.35	26.30	
Drip cap, metal	4.000	L.F.	.080	.84	2.36	3.20	
TOTAL			3.061	286.77	87.65	374.42	

The cost of this system is on a cost per each window basis.

Awning Window System Components

Component Description	QUAN.	UNIT	LABOR-HOURS	COST EACH		
				MAT.	INST.	TOTAL
Windows, awning						
Windows, awning, builder's quality, 34" x 22", single glass	1.000	Ea.	.800	187.00	23.50	210.50
40" x 28", single glass	1.000	Ea.	.889	242.00	26.00	268.00
Plastic clad premium insulating glass, 34" x 22"	1.000	Ea.	.800	193.00	23.50	216.50
40" x 22"	1.000	Ea.	.800	211.00	23.50	234.50
36" x 36"	1.000	Ea.	.889	310.00	26.00	336.00
48" x 28"	1.000	Ea.	1.000	269.00	29.50	298.50
Metal clad deluxe insulating glass, 34" x 22"	1.000	Ea.	.800	175.00	23.50	198.50
40" x 22"	1.000	Ea.	.800	191.00	23.50	214.50
36" x 25"	1.000	Ea.	.889	184.00	26.00	210.00
40" x 30"	1.000	Ea.	.889	209.00	26.00	235.00

Sliding Window Systems

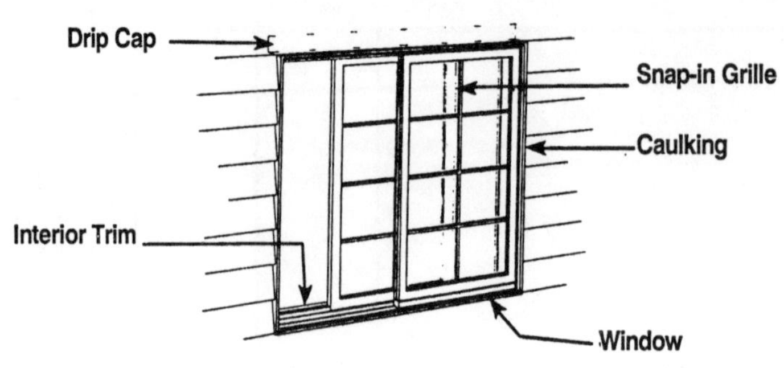

System Description	QUAN.	UNIT	LABOR-HOURS	COST EACH			EA. EXTENSIONS
				MAT.	INST.	TOTAL	
BUILDER'S QUALITY WOOD WINDOW, 3' X 2', SLIDING							
Window, primed, builder's quality, 3' x 2', insul. glass	1.000	Ea.	.800	177.00	23.50	200.50	
Trim, interior casing	11.000	L.F.	.367	8.69	10.78	19.47	
Paint, interior & exterior, primer & 2 coats	2.000	Face	1.600	1.74	42.00	43.74	
Caulking	10.000	L.F.	.308	.90	9.10	10.00	
Snap-in grille	1.000	Set	.333	28.50	9.80	38.30	
Drip cap, metal	3.000	L.F.	.060	.63	1.77	2.40	
TOTAL			3.468	217.46	96.95	314.41	
PLASTIC CLAD WOOD WINDOW 3' X 3' SLIDING							
Window, premium, 3' x 3', insulating glass	1.000	Ea.	.800	475.00	23.50	498.50	
Trim, interior casing	13.000	L.F.	.433	10.27	12.74	23.01	
Paint, interior, primer & 2 coats	1.000	Face	.800	.87	21.00	21.87	
Caulking	12.000	L.F.	.369	1.08	10.92	12.00	
Snap-in grille	1.000	Set	.333	28.50	9.80	38.30	
TOTAL			2.735	515.72	77.96	593.68	
METAL CLAD WOOD WINDOW 3' X 3' SLIDING							
Window, deluxe, 3' x 3', insulating glass	1.000	Ea.	.800	241.00	23.50	264.50	
Trim, interior casing	13.000	L.F.	.433	10.27	12.74	23.01	
Paint, interior, primer & 2 coats	1.000	Face	.800	.87	21.00	21.87	
Caulking	12.000	L.F.	.369	1.08	10.92	12.00	
Snap-in grille	1.000	Set	.333	28.50	9.80	38.30	
Drip cap, metal	3.000	L.F.	.060	.63	1.77	2.40	
TOTAL			2.795	282.35	79.73	362.08	
BUILDER'S QUALITY WOOD WINDOW 4' X 3'-6" SLIDING							
Window, builder's quality, 4' x 3'-6", insulating glass	1.000	Ea.	.889	247.00	26.00	273.00	
Trim, interior casing	16.000	L.F.	.533	12.64	15.68	28.32	
Paint, interior & exterior, primer & 2 coats	2.000	Face	1.600	1.74	42.00	43.74	
Caulking	15.000	L.F.	.523	1.53	15.47	17.00	
Snap-in grille	1.000	Set	.333	28.50	9.80	38.30	
Drip cap, metal	4.000	L.F.	.080	.84	2.36	3.20	
TOTAL			3.958	292.25	111.31	403.56	

Sliding Window Systems

System Description	QUAN.	UNIT	LABOR-HOURS	COST EACH			EA. EXTENSIONS
				MAT.	INST.	TOTAL	
PLASTIC CLAD WOOD WINDOW, 4' X 3'-6'', SLIDING							
Window, plastic clad, premium, 4' x 3'-6'', insulating glass	1.000	Ea.	.889	585.00	26.00	611.00	
Trim, interior casing	16.000	L.F.	.533	12.64	15.68	28.32	
Paint, interior, primer & 2 coats	1.000	Face	.800	.87	21.00	21.87	
Caulking	17.000	L.F.	.523	1.53	15.47	17.00	
Snap-in grille	1.000	Set	.333	28.50	9.80	38.30	
TOTAL			3.078	628.54	87.95	716.49	
METAL CLAD WOOD WINDOW 4' X 3'-6'' SLIDING							
Window, deluxe, 4' x 3'-6'', insulating glass	1.000	Ea.	.889	300.00	26.00	326.00	
Trim, interior casing	16.000	L.F.	.533	12.64	15.68	28.32	
Paint, interior, primer & 2 coats	1.000	Face	.800	.87	21.00	21.87	
Caulking	15.000	L.F.	.523	1.53	15.47	17.00	
Snap-in grille	1.000	Set	.333	28.50	9.80	38.30	
Drip cap, metal	4.000	L.F.	.080	.84	2.36	3.20	
TOTAL			3.158	344.38	90.31	434.69	
BUILDER'S QUALITY WOOD WINDOW 6' X 5' SLIDING							
Window, builder's quality, 6' x 5', insulating glass	1.000	Ea.	1.000	405.00	29.50	434.50	
Trim, interior casing	23.000	L.F.	.767	18.17	22.54	40.71	
Paint, interior & exterior, primer & 2 coats	2.000	Face	1.600	1.74	42.00	43.74	
Caulking	22.000	L.F.	.677	1.98	20.02	22.00	
Snap-in grille	1.000	Set	.364	47.00	10.65	57.65	
Drip cap, metal	6.000	L.F.	.120	1.26	3.54	4.80	
TOTAL			4.528	475.15	128.25	603.40	
METAL CLAD WOOD WINDOW, 6' X 5', SLIDING							
Window, metal clad, deluxe, 6' x 5', insulating glass	1.000	Ea.	1.000	445.00	29.50	474.50	
Trim, interior casing	23.000	L.F.	.767	18.17	22.54	40.71	
Paint, interior, primer & 2 coats	1.000	Face	.800	.87	21.00	21.87	
Caulking	22.000	L.F.	.677	1.98	20.02	22.00	
Snap-in grille	1.000	Set	.364	47.00	10.65	57.65	
Drip cap, metal	6.000	L.F.	.120	1.26	3.54	4.80	
TOTAL			3.728	514.28	107.25	621.53	

The cost of this system is on a cost per each window basis.

Sliding Window System Components

Component Description	QUAN.	UNIT	LABOR-HOURS	COST EACH		
				MAT.	INST.	TOTAL
Windows, sliding						
Windows, sliding, builder's quality, 3' x 2', single glass	1.000	Ea	1.000	520.00	29.50	549.50
6' x 5', single glass	1.000	Ea.	1.000	252.00	29.50	281.50
Plastic clad premium insulating glass, 3' x 3'	1.000	Ea.	.800	475.00	23.50	498.50
5' x 4'	1.000	Ea.	.889	680.00	26.00	706.00
6' x 5'	1.000	Ea.	1.000	845.00	29.50	874.50
Metal clad deluxe insulating glass, 3' x 3'	1.000	Ea.	.800	241.00	23.50	264.50
5' x 4'	1.000	Ea.	.889	365.00	26.00	391.00

Bow/Bay Window Systems

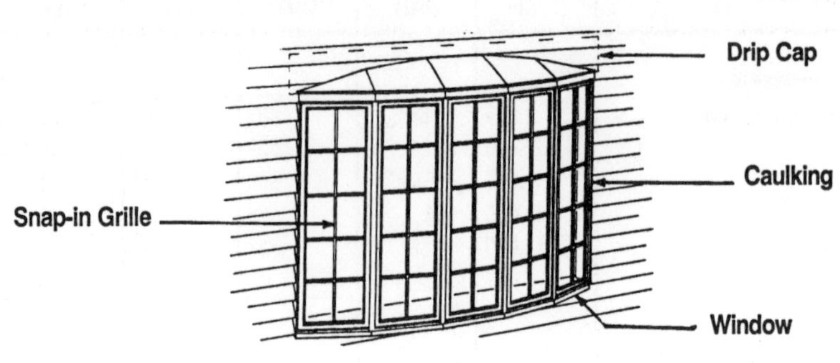

System Description	QUAN.	UNIT	LABOR-HOURS	COST EACH			EA. EXTENSIONS
				MAT.	INST.	TOTAL	
AWNING TYPE BOW WINDOW, BUILDER'S QUALITY, 8' X 5'							
Window, primed, builder's quality, 8' x 5', insulating glass	1.000	Ea.	1.600	945.00	47.00	992.00	
Trim, interior casing	27.000	L.F.	.900	21.33	26.46	47.79	
Paint, interior & exterior, primer & 1 coat	2.000	Face	3.200	9.26	84.00	93.26	
Drip cap, vinyl	1.000	Ea.	.533	71.50	15.65	87.15	
Caulking	26.000	L.F.	.800	2.34	23.66	26.00	
Snap-in grilles	1.000	Set	1.067	55.60	31.40	87.00	
TOTAL			8.100	1105.03	228.17	1333.20	
AWNING TYPE BOW WINDOW PLASTIC CLAD 10' X 5'							
Window, premium, 10' x 5', insulating glass	1.000	Ea.	2.286	2175.00	67.00	2242.00	
Trim, interior casing	33.000	L.F.	1.100	26.07	32.34	58.41	
Paint, interior, primer & 2 coats	1.000	Face	1.600	1.74	42.00	43.74	
Drip cap, vinyl	1.000	Ea.	.615	78.00	18.05	96.05	
Caulking	32.000	L.F.	.985	2.88	29.12	32.00	
Snap-in grille	5.000	Set	1.333	69.50	39.25	108.75	
TOTAL			7.919	2353.19	227.76	2580.95	
CASEMENT TYPE BOW WINDOW BUILDER'S QUALITY 12' X 6'							
Window, builder's quality, 12' x 6', insulating glass	1.000	Ea.	2.667	2800.00	78.50	2878.50	
Trim, interior casing	37.000	L.F.	1.233	29.23	36.26	65.49	
Paint, interior & exterior, primer & 2 coats	2.000	Face	2.667	6.18	70.00	76.18	
Drip cap, vinyl	1.000	Ea.	.615	78.00	18.05	96.05	
Caulking	36.000	L.F.	1.108	3.24	32.76	36.00	
Snap-in grille	6.000	Set	1.600	83.40	47.10	130.50	
TOTAL			9.890	3000.05	282.67	3282.72	
CASEMENT TYPE BOW WINDOW, PLASTIC CLAD, 10' X 6'							
Window, plastic clad, premium, 10' x 6', insulating glass	1.000	Ea.	2.286	2075.00	67.00	2142.00	
Trim, interior casing	33.000	L.F.	1.100	26.07	32.34	58.41	
Paint, interior, primer & 1 coat	1.000	Face	1.600	1.74	42.00	43.74	
Drip cap, vinyl	1.000	Ea.	.615	78.00	18.05	96.05	
Caulking	32.000	L.F.	.985	2.88	29.12	32.00	
Snap-in grilles	1.000	Set	1.333	69.50	39.25	108.75	
TOTAL			7.919	2253.19	227.76	2480.95	

Bow/Bay Window Systems

System Description	QUAN.	UNIT	LABOR-HOURS	COST EACH			EA. EXTENSIONS
				MAT.	INST.	TOTAL	
CASEMENT TYPE BOW WINDOW METAL CLAD 12' X 6'							
Window, deluxe, 12' x 6', insulating glass	1.000	Ea.	2.667	2175.00	78.50	2253.50	
Trim, interior casing	37.000	L.F.	1.233	29.23	36.26	65.49	
Paint, interior, primer & 2 coats	1.000	Face	2.667	6.18	70.00	76.18	
Drip cap, vinyl	1.000	Ea.	.615	78.00	18.05	96.05	
Caulking	36.000	L.F.	1.108	3.24	32.76	36.00	
Snap-in grille	6.000	Set	1.600	83.40	47.10	130.50	
TOTAL			9.890	2375.05	282.67	2657.72	
DOUBLE HUNG TYPE BOW WINDOW BUILDER'S QUALITY 9' X 5'							
Window, builder's quality 9' x 5' insulating glass	1.000	Ea.	1.600	1225.00	47.00	1272.00	
Trim interior casing	27.000	L.F.	.967	22.91	28.42	51.33	
Paint interior & exterior primer & 2 coats	2.000	Face	3.200	3.48	84.00	87.48	
Drip cap vinyl	1.000	Ea.	.615	78.00	18.05	96.05	
Caulking	26.000	L.F.	.862	2.52	25.48	28.00	
Snap-in grille	4.000	Set	1.067	55.60	31.40	87.00	
TOTAL			8.311	1387.51	234.35	1621.86	
DOUBLE HUNG TYPE BOW WINDOW PLASTIC CLAD 9' X 5'							
Window, premium 9' x 5' insulating glass	1.000	Ea.	2.667	1525.00	78.50	1603.50	
Trim interior casing	27.000	L.F.	.967	22.91	28.42	51.33	
Paint interior primer & 2 coats	1.000	Face	1.600	1.74	42.00	43.74	
Drip cap vinyl	1.000	Ea.	.615	78.00	18.05	96.05	
Caulking	26.000	L.F.	.862	2.52	25.48	28.00	
Snap-in grille	4.000	Set	1.067	55.60	31.40	87.00	
TOTAL			7.778	1685.77	223.85	1909.62	
DOUBLE HUNG TYPE, METAL CLAD, 9' X 5'							
Window, metal clad, deluxe, 9' x 5', insulating glass	1.000	Ea.	2.667	1075.00	78.50	1153.50	
Trim, interior casing	29.000	L.F.	.967	22.91	28.42	51.33	
Paint, interior, primer & 1 coat	1.000	Face	1.600	1.74	42.00	43.74	
Drip cap, vinyl	1.000	Set	.615	78.00	18.05	96.05	
Caulking	28.000	L.F.	.862	2.52	25.48	28.00	
Snap-in grilles	1.000	Set	1.067	55.60	31.40	87.00	
TOTAL			7.778	1235.77	223.85	1459.62	

The cost of this system is on a cost per each window basis.

Bow/Bay Window System Components

Component Description	QUAN.	UNIT	LABOR-HOURS	COST EACH		
				MAT.	INST.	TOTAL
Windows						
Windows, bow awning type, builder's quality, 8' x 5', single glass	1.000	Ea.	1.600	820.00	47.00	867.00
Insulating glass	1.000	Ea.	1.600	945.00	47.00	992.00
Plastic clad premium insulating glass, 6' x 4'	1.000	Ea.	1.600	1275.00	47.00	1322.00
Metal clad deluxe insulating glass, 6' x 4'	1.000	Ea.	1.600	795.00	47.00	842.00
Bow casement type, builder's quality, 8' x 5', single glass	1.000	Ea.	1.600	1175.00	47.00	1222.00
Insulating glass	1.000	Ea.	1.600	1175.00	47.00	1222.00
Plastic clad premium insulating glass, 8' x 5'	1.000	Ea.	1.600	1375.00	47.00	1422.00
Metal clad deluxe insulating glass, 8' x 5'	1.000	Ea.	1.600	1400.00	47.00	1447.00
Bow, double hung type, builder's quality, 8' x 4', single glass	1.000	Ea.	1.600	1225.00	47.00	1272.00
Insulating glass	1.000	Ea.	1.600	1225.00	47.00	1272.00

Fixed Window Systems

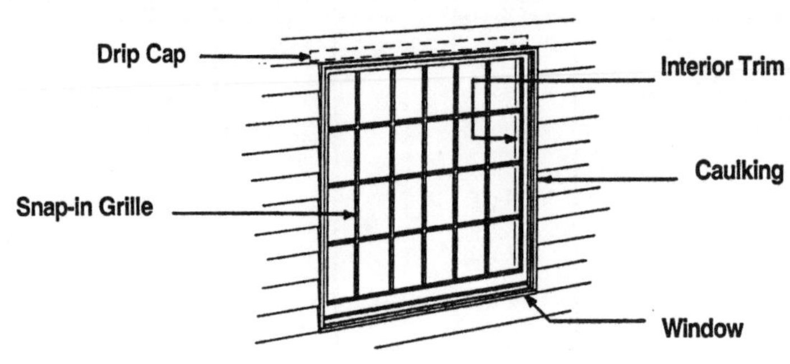

System Description	QUAN.	UNIT	LABOR-HOURS	COST EACH			EA. EXTENSIONS
				MAT.	INST.	TOTAL	
BUILDER'S QUALITY PICTURE WINDOW, 4' X 4'							
Window, primed, builder's quality, 4' x 4', insulating glass	1.000	Ea.	1.333	216.00	39.00	255.00	
Trim, interior casing	17.000	L.F.	.567	13.43	16.66	30.09	
Paint, interior & exterior, primer & 2 coats	2.000	Face	1.600	1.74	42.00	43.74	
Caulking	16.000	L.F.	.492	1.44	14.56	16.00	
Snap-in grille	1.000	Ea.	.267	84.50	7.85	92.35	
Drip cap, metal	4.000	L.F.	.080	.84	2.36	3.20	
TOTAL			4.339	317.95	122.43	440.38	
PLASTIC CLAD WOOD WINDOW, 4' X 4'							
Window, premium, 4' x 4', insulating glass	1.000	Ea.	1.333	405.00	39.00	444.00	
Trim, interior casing	17.000	L.F.	.567	13.43	16.66	30.09	
Paint, interior, primer & 2 coats	1.000	Face	.800	.87	21.00	21.87	
Caulking	16.000	L.F.	.492	1.44	14.56	16.00	
Snap-in grille	1.000	Ea.	.267	84.50	7.85	92.35	
Drip cap, metal	4.000	L.F.	.080	.84	2.36	3.20	
TOTAL			3.539	506.08	101.43	607.51	
METAL CLAD WOOD WINDOW, 4' X 4'							
Window, deluxe, 4' x 4', insulating glass	1.000	Ea.	1.333	267.00	39.00	306.00	
Trim, interior casing	17.000	L.F.	.567	13.43	16.66	30.09	
Paint, interior, primer & 2 coats	1.000	Face	.800	.87	21.00	21.87	
Caulking	16.000	L.F.	.492	1.44	14.56	16.00	
Snap-in grille	1.000	Ea.	.267	84.50	7.85	92.35	
Drip cap, metal	4.000	L.F.	.080	.84	2.36	3.20	
TOTAL			3.539	368.08	101.43	469.51	
BUILDER'S QUALITY PICTURE WINDOW, 6' X 4'-6"							
Window, builder's quality, 6' x 4'-6", insulating glass	1.000	Ea.	1.600	350.00	47.00	397.00	
Trim, interior casing	23.000	L.F.	.767	18.17	22.54	40.71	
Paint, interior & exterior, primer & 2 coats	2.000	Face	1.600	1.74	42.00	43.74	
Caulking	22.000	L.F.	.677	1.98	20.02	22.00	
Snap-in grille	1.000	Ea.	.286	111.00	8.40	119.40	
Drip cap, metal	6.000	L.F.	.120	1.26	3.54	4.80	
TOTAL			5.050	484.15	143.50	627.65	

Fixed Window Systems

System Description	QUAN.	UNIT	LABOR-HOURS	COST EACH			EA. EXTENSIONS
				MAT.	INST.	TOTAL	
PLASTIC CLAD WOOD WINDOW, 4'-6'' X 6'-6''							
Window, plastic clad, prem., 4'-6'' x 6'-6'', insul. glass	1.000	Ea.	1.455	585.00	42.50	627.50	
Trim, interior casing	23.000	L.F.	.767	18.17	22.54	40.71	
Paint, interior, primer & 2 coats	1.000	Face	.800	.87	21.00	21.87	
Caulking	22.000	L.F.	.677	1.98	20.02	22.00	
Snap-in grille	1.000	Ea.	.267	84.50	7.85	92.35	
TOTAL			3.966	690.52	113.91	804.43	
METAL CLAD WOOD WINDOW, 4'-6'' X 6'-6''							
Window, deluxe, 4'-6'' x 6'-6'', insulating glass	1.000	Ea.	1.455	395.00	42.50	437.50	
Trim, interior casing	23.000	L.F.	.767	18.17	22.54	40.71	
Paint, interior, primer & 2 coats	1.000	Face	.800	.87	21.00	21.87	
Caulking	22.000	L.F.	.677	1.98	20.02	22.00	
Snap-in grille	1.000	Ea.	.267	84.50	7.85	92.35	
Drip cap, metal	4.500	L.F.	.090	.95	2.66	3.61	
TOTAL			4.056	501.47	116.57	618.04	
PLASTIC CLAD, WOOD, WINDOW, 6'-6'' X 6'-6''							
Window, premium, 6'-6'' x 6'-6'', insulating glass	1.000	Ea.	1.600	1075.00	47.00	1122.00	
Trim, interior casing	27.000	L.F.	.900	21.33	26.46	47.79	
Paint, interior, primer & 2 coats	1.000	Face	1.600	4.63	42.00	46.63	
Caulking	26.000	L.F.	.800	2.34	23.66	26.00	
Snap-in grille	1.000	Ea.	.267	84.50	7.85	92.35	
Drip cap, metal	6.500	L.F.	.130	1.37	3.84	5.21	
TOTAL			5.297	1189.17	150.81	1339.98	
METAL CLAD WOOD WINDOW, 6'-6'' X 6'-6''							
Window, metal clad, deluxe, 6'-6'' x 6'-6'', insulating glass	1.000	Ea.	1.600	605.00	47.00	652.00	
Trim interior casing	27.000	L.F.	.900	21.33	26.46	47.79	
Paint, interior, primer & 2 coats	1.000	Face	1.600	4.63	42.00	46.63	
Caulking	26.000	L.F.	.800	2.34	23.66	26.00	
Snap-in grille	1.000	Ea.	.267	84.50	7.85	92.35	
Drip cap, metal	6.500	L.F.	.130	1.37	3.84	5.21	
TOTAL			5.297	719.17	150.81	869.98	

The cost of this system is on a cost per each window basis.

Fixed Window System Components

Component Description	QUAN.	UNIT	LABOR-HOURS	COST EACH		
				MAT.	INST.	TOTAL
Window, picture						
Window-picture, builder's quality, 4' x 4', single glass	1.000	Ea.	1.333	191.00	39.00	230.00
6' x 4'-6'', single glass	1.000	Ea.	1.600	345.00	47.00	392.00
Plastic clad premium insulating glass, 4' x 4'	1.000	Ea.	1.333	405.00	39.00	444.00
5'-6'' x 6'-6''	1.000	Ea.	1.600	760.00	47.00	807.00
Metal clad deluxe insulating glass, 4' x 4'	1.000	Ea.	1.333	267.00	39.00	306.00
5'-6'' x 6'-6''	1.000	Ea.	1.600	495.00	47.00	542.00
Trim						
Trim, interior casing, window 4' x 4'	17.000	L.F.	.567	13.45	16.65	30.10
4'-6'' x 4'-6''	19.000	L.F.	.633	15.00	18.60	33.60

Entrance Door Systems

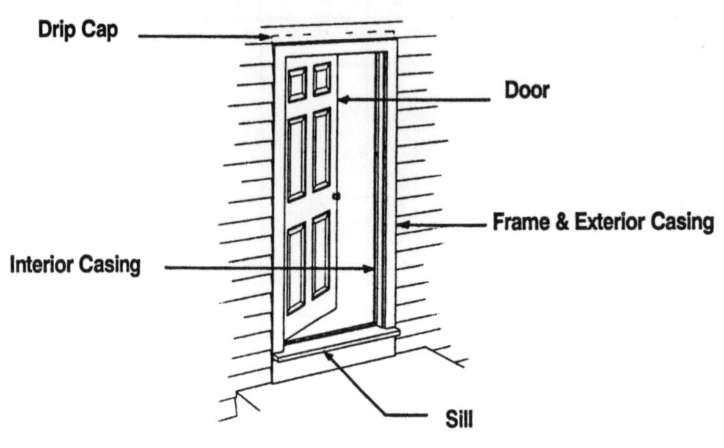

System Description	QUAN.	UNIT	LABOR-HOURS	COST EACH			EA. EXTENSIONS
				MAT.	INST.	TOTAL	
COLONIAL, 6 PANEL, 3' X 6'-8'', WOOD							
Door, 3' x 6'-8'' x 1-3/4'' thick, pine, 6 panel colonial	1.000	Ea.	1.067	355.00	28.85	383.85	
Frame, 5-13/16'' deep, incl. exterior casing & drip cap	17.000	L.F.	.725	85.85	19.38	105.23	
Interior casing, 2-1/2'' wide	18.000	L.F.	.600	14.22	17.64	31.86	
Sill, 8/4 x 8'' deep	3.000	L.F.	.480	36.30	12.90	49.20	
Butt hinges, brass, 4-1/2'' x 4-1/2''	1.500	Pr.		13.28		13.28	
Lockset	1.000	Ea.	.571	28.50	16.75	45.25	
Weatherstripping, metal, spring type, bronze	1.000	Set	1.053	8.35	31.00	39.35	
Paint, interior & exterior, primer & 2 coats	2.000	Face	1.778	9.16	47.00	56.16	
TOTAL			6.274	550.66	173.52	724.18	
SOLID CORE BIRCH, FLUSH, 3' X 6'-8''							
Door, 3' x 6'-8'', 1-3/4'' thick, birch, flush solid core	1.000	Ea.	1.067	105.00	28.85	133.85	
Frame, 5-13/16'' deep, incl. exterior casing & drip cap	17.000	L.F.	.725	85.85	19.38	105.23	
Interior casing, 2-1/2'' wide	18.000	L.F.	.600	14.22	17.64	31.86	
Sill, 8/4 x 8'' deep	3.000	L.F.	.480	36.30	12.90	49.20	
Butt hinges, brass, 4-1/2'' x 4-1/2''	1.500	Pr.		13.28		13.28	
Lockset	1.000	Ea.	.571	28.50	16.75	45.25	
Weatherstripping, metal, spring type, bronze	1.000	Set	1.053	8.35	31.00	39.35	
Paint, interior & exterior, primer & 2 coats	2.000	Face	1.778	8.58	47.00	55.58	
TOTAL			6.274	300.08	173.52	473.60	
DUTCH, PINE, 3'-0'' X 6'-8''							
Door, 3'-0'' x 6'-8'' x 1-3/4'' thick, pine	1.000	Ea.	1.600	775.00	43.00	818.00	
Frame, 5-13/16'' deep, exterior casing & drip cap	17.000	L.F.	.725	86.00	19.40	105.40	
Interior casing, 2-1/2'' wide	19.000	L.F.	.600	14.20	17.65	31.85	
Sill, 8/4 x 8'' deep	3.000	L.F.	.480	36.50	12.90	49.40	
Butt hinges, brass, 4-1/2'' x 4-1/2''	1.500	Pr.		13.30		13.30	
Lockset	1.000	Ea.	.571	28.50	16.75	45.25	
Weatherstripping, metal, spring type, bronze	1.000	Set	1.053	8.35	31.00	39.35	
Paint, Interior & exterior, primer & 2 coats	2.000	Face	1.778	8.60	47.00	55.60	
TOTAL			6.807	970.00	187.00	1157.00	

Entrance Door Systems

System Description	QUAN.	UNIT	LABOR-HOURS	COST EACH			EA. EXTENSIONS
				MAT.	INST.	TOTAL	
COLONIAL, 8 PANEL, 3'-0'' X 6'-8''							
Door, 3'-0'' x 6'-8'' x 1-3/4'' thick, pine	1.000	Ea.	1.067	480.00	29.00	509.00	
Frame, 5-13/16'' deep, exterior casing & drip cap	17.000	L.F.	.725	86.00	19.40	105.40	
Interior casing, 2-1/2'' wide	19.000	L.F.	.600	14.20	17.65	31.85	
Sill, 8/4 x 8'' deep	3.000	L.F.	.480	36.50	12.90	49.40	
Butt hinges, brass, 4-1/2'' x 4-1/2''	1.500	Pr.		13.30		13.30	
Lockset	1.000	Ea.	.571	28.50	16.75	45.25	
Weatherstripping, metal, spring type, bronze	1.000	Set	1.053	8.35	31.00	39.35	
Paint, Interior & exterior, primer & 2 coats	2.000	Face	1.778	8.60	47.00	55.60	
TOTAL			6.274	675.00	174.00	849.00	
METAL CLAD WOOD, RAISED PANEL, 3'-0'' X 6'-8''							
Door, 3'-0'' x 6'-8'' x 1-3/8'' thick, metal clad wood	1.000	Ea.	1.067	202.00	29.00	231.00	
Frame, 5-13/16'' deep, exterior casing & drip cap	17.000	L.F.	.725	86.00	19.40	105.40	
Interior casing, 2-1/2'' wide	19.000	L.F.	.600	14.20	17.65	31.85	
Sill, 8/4 x 8'' deep	3.000	L.F.	.480	36.50	12.90	49.40	
Butt hinges, brass, 4-1/2'' x 4-1/2''	1.500	Pr.		13.30		13.30	
Lockset	1.000	Ea.	.571	28.50	16.75	45.25	
Weatherstripping, metal, spring type, bronze	1.000	Set	1.053	8.35	31.00	39.35	
Paint, Interior & exterior, primer & 2 coats	2.000	Face	1.778	8.60	47.00	55.60	
TOTAL			6.274	395.00	174.00	569.00	
DELUXE METAL, RAISED PANEL, 3'-0'' X 6'-8''							
Door, 3'-0'' x 6'-8'' x 1-3/8'' thick, deluxe metal	1.000	Ea.	1.231	300.00	33.00	333.00	
Frame, 5-13/16'' deep, exterior casing & drip cap	17.000	L.F.	.725	86.00	19.40	105.40	
Interior casing, 2-1/2'' wide	19.000	L.F.	.600	14.20	17.65	31.85	
Sill, 8/4 x 8'' deep	3.000	L.F.	.480	36.50	12.90	49.40	
Butt hinges, brass, 4-1/2'' x 4-1/2''	1.500	Pr.		13.30		13.30	
Lockset	1.000	Ea.	.571	28.50	16.75	45.25	
Weatherstripping, metal, spring type, bronze	1.000	Set	1.053	8.35	31.00	39.35	
Paint, interior & exterior, primer & 2 coats	2.000	Face	1.778	8.60	47.00	55.60	
TOTAL			6.438	495.00	178.00	673.00	

These systems are on a cost per each door basis.

Entrance Door System Components

Component Description	QUAN.	UNIT	LABOR-HOURS	COST EACH		
				MAT.	INST.	TOTAL
Doors, exterior						
Door exterior wood 1-3/4'' thick, pine, dutch door, 2'-8'' x 6'-8'' minimum	1.000	Ea.	1.333	705.00	35.50	740.50
Maximum	1.000	Ea.	1.600	735.00	43.00	778.00
Hand carved mahogany, 2'-8'' x 6'-8''	1.000	Ea.	1.067	460.00	29.00	489.00
3'-0'' x 6'-8''	1.000	Ea.	1.067	495.00	29.00	524.00
Door, metal clad wood 1-3/8'' thick raised panel, 2'-8'' x 6'-8''	1.000	Ea.	1.067	202.00	29.00	231.00
Deluxe metal door, 2'-8'' x 6'-8''	1.000	Ea.	1.231	289.00	33.00	322.00
Butt hinges						
Butt hinges, steel plated, 4-1/2'' x 4-1/2'', plain	1.500	Pr.		13.30		13.30
Ball bearing	1.500	Pr.		29.50		29.50
Lockset						
Lockset, minimum	1.000	Ea.	.571	28.50	16.75	45.25
Maximum	1.000	Ea.	1.000	117.00	29.50	146.50

Sliding Door Systems

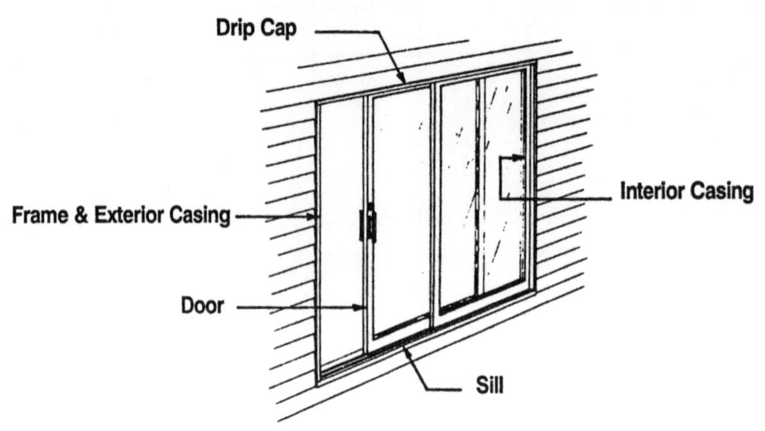

System Description	QUAN.	UNIT	LABOR-HOURS	COST EACH			EA. EXTENSIONS
				MAT.	INST.	TOTAL	
WOOD SLIDING DOOR, 6' WIDE, PREMIUM							
Wood, 5/8'' thick tempered insul. glass, 6' wide	1.000	Ea.	4.000	1025.00	117.00	1142.00	
Interior casing	20.000	L.F.	.667	15.80	19.60	35.40	
Exterior casing	20.000	L.F.	.667	15.80	19.60	35.40	
Sill, oak, 8/4 x 8'' deep	6.000	L.F.	.960	72.60	25.80	98.40	
Drip cap	6.000	L.F.	.120	1.26	3.54	4.80	
Paint, interior & exterior, primer & 2 coats	2.000	Face	2.560	12.00	67.20	79.20	
TOTAL			8.974	1142.46	252.74	1395.20	
WOOD SLIDING DOOR, 8' WIDE, PREMIUM							
Wood, 5/8'' thick tempered insul. glass, 8' wide, premium	1.000	Ea.	5.333	1375.00	157.00	1532.00	
Interior casing	22.000	L.F.	.733	17.38	21.56	38.94	
Exterior casing	22.000	L.F.	.733	17.38	21.56	38.94	
Sill, oak, 8/4 x 8'' deep	8.000	L.F.	1.280	96.80	34.40	131.20	
Drip cap	8.000	L.F.	.160	1.68	4.72	6.40	
Paint, interior & exterior, primer & 2 coats	2.000	Face	2.816	13.20	73.92	87.12	
TOTAL			11.055	1521.44	313.16	1834.60	
WOOD SLIDING DOOR, 12' WIDE, PREMIUM							
Wood, 5/8'' thick tempered insul. glass, 12' wide	1.000	Ea.	6.400	2125.00	188.00	2313.00	
Interior casing	26.000	L.F.	.867	20.54	25.48	46.02	
Exterior casing	26.000	L.F.	.867	20.54	25.48	46.02	
Sill, oak, 8/4 x 8'' deep	12.000	L.F.	1.920	145.20	51.60	196.80	
Drip cap	12.000	L.F.	.240	2.52	7.08	9.60	
Paint, interior & exterior, primer & 2 coats	2.000	Face	3.328	15.60	87.36	102.96	
TOTAL			13.622	2329.40	385.00	2714.40	
ALUMINUM SLIDING DOOR, 6' WIDE, PREMIUM							
Aluminum, 5/8'' thick tempered insul. glass, 6' wide	1.000	Ea.	4.000	425.00	117.00	542.00	
Interior casing	20.000	L.F.	.667	15.80	19.60	35.40	
Exterior casing	20.000	L.F.	.667	15.80	19.60	35.40	
Sill, oak, 8/4 x 8'' deep	6.000	L.F.	.960	72.60	25.80	98.40	
Drip cap	6.000	L.F.	.120	1.26	3.54	4.80	
Paint, interior & exterior, primer & 2 coats	2.000	Face	2.560	12.00	67.20	79.20	
TOTAL			8.974	542.46	252.74	795.20	

Sliding Door Systems

System Description	QUAN.	UNIT	LABOR-HOURS	COST EACH			EA. EXTENSIONS
				MAT.	INST.	TOTAL	
ALUMINUM SLIDING DOOR, 8' WIDE, PREMIUM							
Aluminum, 5/8'' tempered insul. glass, 8' wide, premium	1.000	Ea.	5.333	515.00	157.00	672.00	
Interior casing	22.000	L.F.	.733	17.38	21.56	38.94	
Exterior casing	22.000	L.F.	.733	17.38	21.56	38.94	
Sill, oak, 8/4 x 8'' deep	8.000	L.F.	1.280	96.80	34.40	131.20	
Drip cap	8.000	L.F.	.160	1.68	4.72	6.40	
Paint, interior & exterior, primer & 2 coats	2.000	Face	2.816	13.20	73.92	87.12	
TOTAL			11.055	661.44	313.16	974.60	
ALUMINUM SLIDING DOOR, 12' WIDE, PREMIUM							
Aluminum, 5/8'' thick tempered insul. glass, 12' wide	1.000	Ea.	6.400	810.00	188.00	998.00	
Interior casing	26.000	L.F.	.867	20.50	25.50	46.00	
Exterior casing	26.000	L.F.	.867	20.50	25.50	46.00	
Sill, oak, 8/4 x 8'' deep	12.000	L.F.	1.920	145.00	51.50	196.50	
Drip cap	12.000	L.F.	.240	2.52	7.10	9.62	
Paint, interior & exterior, primer & 2 coats	2.000	Face	3.328	15.60	87.50	103.10	
TOTAL			13.622	1025.00	385.00	1410.00	

The cost of this system is on a cost per each door basis.

Sliding Door System Components

Component Description	QUAN.	UNIT	LABOR-HOURS	COST EACH		
				MAT.	INST.	TOTAL
Sliding doors						
Sliding door, wood, 5/8'' thick, tempered insul. glass, 6' wide, premium	1.000	Ea.	4.000	1025.00	117.00	1142.00
Economy	1.000	Ea.	4.000	780.00	117.00	897.00
8'wide, wood premium	1.000	Ea.	5.333	1375.00	157.00	1532.00
Economy	1.000	Ea.	5.333	1050.00	157.00	1207.00
12' wide, wood premium	1.000	Ea.	6.400	2125.00	188.00	2313.00
Economy	1.000	Ea.	6.400	1625.00	188.00	1813.00
Aluminum, 5/8'' thick, tempered insul. glass, 6'wide, premium	1.000	Ea.	4.000	425.00	117.00	542.00
Economy	1.000	Ea.	4.000	288.00	117.00	405.00
8'wide, premium	1.000	Ea.	5.333	515.00	157.00	672.00
Economy	1.000	Ea.	5.333	350.00	157.00	507.00
12' wide, premium	1.000	Ea.	6.400	810.00	188.00	998.00
Economy	1.000	Ea.	6.400	550.00	188.00	738.00
Sills						
Sill, oak, 8/4 x 8'' deep, 6' wide door	6.000	L.F.	.960	72.50	25.50	98.00
8' wide door	8.000	L.F.	1.280	97.00	34.50	131.50
12' wide door	12.000	L.F.	1.920	145.00	51.50	196.50
8/4 x 10'' deep, 6' wide door	6.000	L.F.	1.067	98.00	29.00	127.00
8' wide door	8.000	L.F.	1.422	130.00	38.50	168.50
12' wide door	12.000	L.F.	2.133	196.00	57.50	253.50
Paint or Stain						
Paint or stain, interior & exterior, 6' wide door, 1 coat	2.000	Face	1.600	4.00	42.50	46.50
8' wide door, 1 coat	2.000	Face	1.760	4.40	46.50	50.90
12' wide door, 1 coat	2.000	Face	2.080	5.20	55.00	60.20
Aluminum door, trim only, interior & exterior, 6' door, 1 coat	2.000	Face	.800	2.00	21.00	23.00
8' wide door, 1 coat	2.000	Face	.880	2.20	23.50	25.70
12' wide door, 1 coat	2.000	Face	1.040	2.60	27.50	30.10

Residential Garage Door Systems

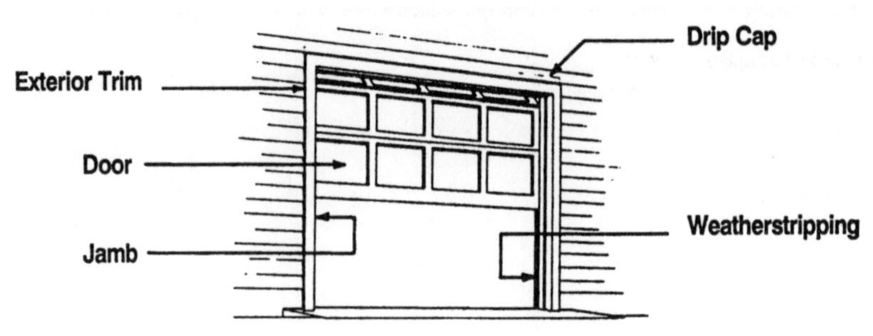

System Description	QUAN.	UNIT	LABOR-HOURS	COST EACH			EA. EXTENSIONS
				MAT.	INST.	TOTAL	
OVERHEAD, SECTIONAL GARAGE DOOR, 9' X 7'							
Wood, overhead sectional door, std., incl. hardware, 9' x 7'	1.000	Ea.	2.000	340.00	58.50	398.50	
Jamb & header blocking, 2'' x 6''	25.000	L.F.	.901	14.50	27.75	42.25	
Exterior trim	25.000	L.F.	.833	19.75	24.50	44.25	
Paint, interior & exterior, primer & 2 coats	2.000	Face	3.556	18.32	94.00	112.32	
Weatherstripping, molding type	1.000	Set	.767	18.17	22.54	40.71	
Drip cap	9.000	L.F.	.180	1.89	5.31	7.20	
TOTAL			8.237	412.63	232.60	645.23	
OVERHEAD, METAL, SECTIONAL GARAGE DOOR, 9' X 7'							
Metal, overhead door, standard, incl. hardware, 9' x 7'	1.000	Ea.	2.000	325.00	58.50	383.50	
Jamb & header blocking, 2'' x 6''	25.000	L.F.	.901	14.50	27.75	42.25	
Exterior trim	25.000	L.F.	.833	19.75	24.50	44.25	
Paint, interior & exterior, primer & 2 coats	2.000	Face	3.556	18.32	94.00	112.32	
Weatherstripping, molding type	1.000	Set	.767	18.17	22.54	40.71	
Drip cap	9.000	L.F.	.180	1.89	5.31	7.20	
TOTAL			8.237	397.63	232.60	630.23	
OVERHEAD, SWING-UP TYPE, GARAGE DOOR, 9' X 7'							
Wood, overhead door, standard, incl. hardware, 9' x 7'	1.000	Ea.	2.000	291.00	58.50	349.50	
Jamb & header blocking, 2'' x 6''	25.000	L.F.	.901	14.50	27.75	42.25	
Exterior trim	25.000	L.F.	.833	19.75	24.50	44.25	
Paint, interior & exterior, primer & 2 coats	2.000	Face	3.556	18.32	94.00	112.32	
Weatherstripping, molding type	1.000	Set	.767	18.17	22.54	40.71	
Drip cap	9.000	L.F.	.180	1.89	5.31	7.20	
TOTAL			8.237	363.63	232.60	596.23	
OVERHEAD, SECTIONAL GARAGE DOOR, 16' X 7'							
Wood, overhead sectional, std., incl. hardware, 16' x 7'	1.000	Ea.	2.667	685.00	78.50	763.50	
Jamb & header blocking, 2'' x 6''	30.000	L.f	1.081	17.40	33.30	50.70	
Exterior trim	30.000	L.F.	1.000	23.70	29.40	53.10	
Paint, interior & exterior, primer & 2 coats	2.000	Face	5.333	27.48	141.00	168.48	
Weatherstripping, molding type	1.000	Set	1.000	23.70	29.40	53.10	
Drip cap	16.000	L.F.	.320	3.36	9.44	12.80	
TOTAL			11.401	780.64	321.04	1101.68	

Residential Garage Door Systems

System Description	QUAN.	UNIT	LABOR-HOURS	COST EACH			EA. EXTENSIONS
				MAT.	INST.	TOTAL	
OVERHEAD, METAL, SECTIONAL GARAGE DOOR, 16' X 7'							
Metal, overhead door, standard, incl. hardware 16' x 7'	1.000	Ea.	2.667	470.00	78.50	548.50	
Jamb & header blocking, 2'' x 6''	30.000	L.F	1.081	17.40	33.30	50.70	
Exterior trim	30.000	L.F.	1.000	23.70	29.40	53.10	
Paint, interior & exterior, primer & 2 coats	2.000	Face	5.333	27.48	141.00	168.48	
Weatherstripping, molding type	1.000	Set	1.000	23.70	29.40	53.10	
Drip cap	16.000	L.F.	.320	3.36	9.44	12.80	
TOTAL			11.401	565.64	321.04	886.68	
OVERHEAD, SWING-UP TYPE, GARAGE DOOR, 16' X 7'							
Wood, overhead, swing-up, std., incl. hardware, 16' x 7'	1.000	Ea.	2.667	505.00	78.50	583.50	
Jamb & header blocking, 2'' x 6''	30.000	L.F.	1.081	17.40	33.30	50.70	
Exterior trim	30.000	L.F.	1.000	23.70	29.40	53.10	
Paint, interior & exterior, primer & 2 coats	2.000	Face	5.333	27.48	141.00	168.48	
Weatherstripping, molding type	1.000	Set	1.000	23.70	29.40	53.10	
Drip cap	16.000	L.F.	.320	3.36	9.44	12.80	
TOTAL			11.401	600.64	321.04	921.68	

This system is on a cost per each door basis.

Residential Garage Door System Components

Component Description	QUAN.	UNIT	LABOR-HOURS	COST EACH		
				MAT.	INST.	TOTAL
Garage doors						
Overhead, sectional, including hardware, fiberglass, 9' x 7', standard	1.000	Ea.	2.000	410.00	58.50	468.50
Deluxe	1.000	Ea.	2.000	475.00	58.50	533.50
16' x 7', standard	1.000	Ea.	2.667	730.00	78.50	808.50
Deluxe	1.000	Ea.	2.667	910.00	78.50	988.50
Hardboard, 9' x 7', standard	1.000	Ea.	2.000	297.00	58.50	355.50
Deluxe	1.000	Ea.	2.000	360.00	58.50	418.50
16' x 7', standard	1.000	Ea.	2.667	530.00	78.50	608.50
Deluxe	1.000	Ea.	2.667	615.00	78.50	693.50
Overhead swing-up type including hardware, fiberglass, 9' x 7', standard	1.000	Ea.	2.000	470.00	58.50	528.50
Deluxe	1.000	Ea.	2.000	555.00	58.50	613.50
16' x 7', standard	1.000	Ea.	2.667	605.00	78.50	683.50
Deluxe	1.000	Ea.	2.667	685.00	78.50	763.50
Hardboard, 9' x 7', standard	1.000	Ea.	2.000	248.00	58.50	306.50
Deluxe	1.000	Ea.	2.000	330.00	58.50	388.50
16' x 7', standard	1.000	Ea.	2.667	345.00	78.50	423.50
Deluxe	1.000	Ea.	2.667	500.00	78.50	578.50
Metal, 9' x 7', standard	1.000	Ea.	2.000	266.00	58.50	324.50
Deluxe	1.000	Ea.	2.000	415.00	58.50	473.50
16' x 7', standard	1.000	Ea.	2.667	415.00	78.50	493.50
Deluxe	1.000	Ea.	2.667	715.00	78.50	793.50
Jamb and header blocking						
Jamb & header blocking, 2'' x 6'', 9' x 7' door	25.000	L.F.	.901	14.50	28.00	42.50
2'' x 8'', 9' x 7' door	25.000	L.F.	1.000	21.00	31.00	52.00
16' x 7' door	30.000	L.F.	1.200	25.00	36.50	61.50
Garage door openers						
Garage door opener, economy	1.000	Ea.	1.000	209.00	29.50	238.50
Deluxe, including remote control	1.000	Ea.	1.000	305.00	29.50	334.50

Aluminum Window Systems

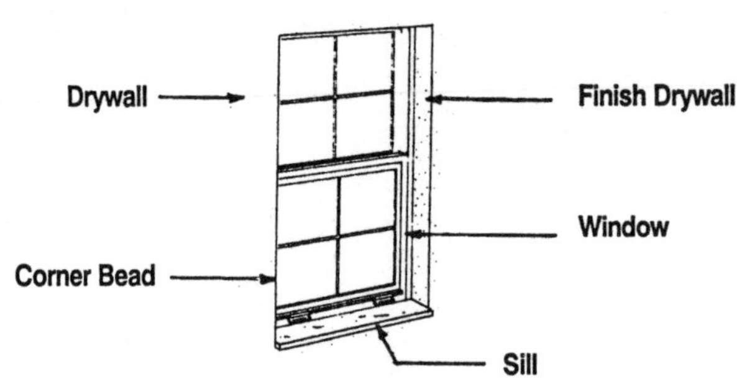

System Description	QUAN.	UNIT	LABOR-HOURS	COST EACH			EA. EXTENSIONS
				MAT.	INST.	TOTAL	
SINGLE HUNG, 2′ X 3′ OPENING							
Window, 2′ x 3′ opening, enameled, insulating glass	1.000	Ea.	1.600	159.00	57.50	216.50	
Blocking, 1″ x 3″ furring strip nailers	10.000	L.F.	.145	3.50	4.30	7.80	
Drywall, 1/2″ thick, standard	5.000	S.F.	.040	.85	1.15	2.00	
Corner bead, 1″ x 1″, galvanized steel	8.000	L.F.	.160	.80	4.72	5.52	
Finish drywall, tape and finish corners inside and outside	16.000	L.F.	.233	1.12	6.88	8.00	
Sill, slate	2.000	L.F.	.400	14.30	10.40	24.70	
TOTAL			2.578	179.57	84.95	264.52	
SLIDING, 3′ X 2′ OPENING							
Window, 3′ x 2′ opening, enameled, insulating glass	1.000	Ea.	1.600	163.00	57.50	220.50	
Blocking, 1″ x 3″ furring strip nailers	10.000	L.F.	.145	3.50	4.30	7.80	
Drywall, 1/2″ thick, standard	5.000	S.F.	.040	.85	1.15	2.00	
Corner bead, 1″ x 1″, galvanized steel	7.000	L.F.	.140	.70	4.13	4.83	
Finish drywall, tape and finish corners inside and outside	14.000	L.F.	.204	.98	6.02	7.00	
Sill, slate	3.000	L.F.	.600	21.45	15.60	37.05	
TOTAL			2.729	190.48	88.70	279.18	
AWNING, 3′-1″ X 3′-2″							
Window, 3′-1″ x 3′-2″ opening, enameled, insul. glass	1.000	Ea.	1.600	182.00	57.50	239.50	
Blocking, 1″ x 3″ furring strip, nailers	12.500	L.F.	.182	4.38	5.38	9.76	
Drywall, 1/2″ thick, standard	4.500	S.F.	.036	.77	1.04	1.81	
Corner bead, 1″ x 1″, galvanized steel	9.250	L.F.	.185	.93	5.46	6.39	
Finish drywall, tape and finish corners, inside and outside	18.500	L.F.	.269	1.30	7.96	9.26	
Sill, slate	3.250	L.F.	.650	23.24	16.90	40.14	
TOTAL			2.922	212.62	94.24	306.86	
CASEMENT, 3′-1″ X 3′-2″							
Window, 3′-1″ x 3′-2″ opening, enameled, insul. glass	1.000	Ea.	1.600	273.00	57.50	330.50	
Blocking, 1″ x 3″ furring strip nailers	12.500	L.F.	.182	4.38	5.40	9.78	
Drywall, 1/2″ thick standard	6.000	S.F.	.048	1.02	1.38	2.40	
Corner bead, 1″ x 1″ galvanized steel	9.000	L.F.	.180	.90	5.30	6.20	
Finish drywall, tape & finish corners, inside & outside	18.000	L.F.	.262	1.26	7.75	9.01	
Sill, slate	3.000	L.F.	.600	21.50	15.60	37.10	
TOTAL			2.872	300.00	93.00	393.00	

Aluminum Window Systems

System Description	QUAN.	UNIT	LABOR-HOURS	COST EACH MAT.	COST EACH INST.	COST EACH TOTAL	EA. EXTENSIONS
SINGLE HUNG, 2'-8'' X 6'-8''							
Window, 2'-8'' x 6'-8'' opening, enameled, insul. glass	1.000	Ea.	2.000	360.00	71.50	431.50	
Blocking, 1'' x 3'' furring strip, nailers	19.000	L.F.	.276	6.65	8.15	14.80	
Drywall, 1/2'' thick standard	9.500	S.F.	.076	1.62	2.19	3.81	
Corner bead, 1'' x 1'' galvanized steel	15.000	L.F.	.300	1.50	8.85	10.35	
Finish drywall, tape & finish corners, inside & outside	30.000	L.F.	.436	2.10	12.90	15.00	
Sill, slate	3.000	L.F.	.600	21.50	15.60	37.10	
TOTAL			3.688	395.00	119.00	514.00	
SLIDING, 5' X 3'							
Window, 5' x 3' opening, enameled, insulating glass	1.000	Ea.	1.778	262.00	63.50	325.50	
Blocking, 1'' x 3'' furring strip, nailers	16.000	L.F.	.233	5.60	6.90	12.50	
Drywall, 1/2'' thick, standard	8.000	L.F.	.064	1.36	1.84	3.20	
Corner bead, 1'' x 1'' galvanized steel	11.000	L.F.	.220	1.10	6.50	7.60	
Finish drywall, tape & finish corners, inside & outside	22.000	L.F.	.320	1.54	9.45	10.99	
Sill, slate	3.000	L.F.	.600	21.50	15.60	37.10	
TOTAL			3.215	293.00	104.00	397.00	
AWNING, 4'-5'' X 5'-3''							
Window, 4'-5'' x 5'-3'' opening, enameled, insul. glass	1.000	Ea.	2.000	315.00	71.50	386.50	
Blocking, 1'' x 3'' furring strip, nailers	20.000	L.F.	.262	6.30	7.75	14.05	
Drywall, 1/2'' thick, standard	10.000	S.F.	.072	1.53	2.07	3.60	
Corner bead, 1'' x 1'' galvanized steel	14.000	L.F.	.260	1.30	7.65	8.95	
Finish drywall, tape & finish corners, inside & outside	28.000	L.F.	.378	1.82	11.20	13.02	
Sill, slate	4.500	L.F.	.800	28.50	21.00	49.50	
TOTAL			3.772	355.00	121.00	476.00	
SINGLE HUNG, 3'-4'' X 5'-0''							
Window, 3'-4'' x 5'-0'' opening, enameled, insul. glass	1.000	Ea.	1.778	179.00	63.50	242.50	
Blocking, 1'' x 3'' furring strip, nailers	16.000	L.F.	.233	5.60	6.90	12.50	
Drywall, 1/2'' thick standard	8.000	S.F.	.064	1.36	1.84	3.20	
Corner bead, 1'' x 1'' galvanized steel	11.000	L.F.	.220	1.10	6.50	7.60	
Finish drywall, tape & finish corners, inside & outside	22.000	L.F.	.320	1.54	9.45	10.99	
Sill, slate	3.500	L.F.	.600	21.50	15.60	37.10	
TOTAL			3.215	210.00	104.00	314.00	

Aluminum Window System Components

Component Description	QUAN.	UNIT	LABOR-HOURS	COST EACH MAT.	COST EACH INST.	COST EACH TOTAL
Windows, aluminum						
Window, aluminum, awning, 3'-1'' x 3'-2'', standard glass	1.000	Ea.	1.600	195.00	57.50	252.50
4'-5'' x 5'-3'', standard glass	1.000	Ea.	2.000	275.00	71.50	346.50
Casement, 3'-1'' x 3'-2'', standard glass	1.000	Ea.	1.600	273.00	57.50	330.50
Single hung, 2' x 3', standard glass	1.000	Ea.	1.600	131.00	57.50	188.50
3'-4'' x 5'-0'', standard glass	1.000	Ea.	1.778	179.00	63.50	242.50
Sliding, 3' x 2', standard glass	1.000	Ea.	1.600	147.00	57.50	204.50
8' x 4', standard glass	1.000	Ea.	2.667	270.00	95.50	365.50
Insulating glass	1.000	Ea.	2.667	435.00	95.50	530.50
Sills						
Wood, 1-5/8'' x 5-1/8'', 2' long	2.000	L.F.	.128	3.96	3.76	7.72
3' long	3.000	L.F.	.192	5.95	5.65	11.60
4' long	4.000	L.F.	.256	7.90	7.50	15.40

Storm Door & Window Systems

Component Description	QUAN.	UNIT	LABOR-HOURS	COST EACH		
				MAT.	INST.	TOTAL
Storm doors						
Storm door, aluminum, combination, storm & screen, anodized, 2'-6'' x 6'-8''	1.000	Ea.	1.067	243.00	29.00	272.00
2'-8'' x 6'-8''	1.000	Ea.	1.143	243.00	31.00	274.00
3'-0'' x 6'-8''	1.000	Ea.	1.143	250.00	31.00	281.00
Mill finish, 2'-6'' x 6'-8''	1.000	Ea.	1.067	233.00	29.00	262.00
2'-8'' x 6'-8''	1.000	Ea.	1.143	238.00	31.00	269.00
3'-0'' x 6'-8''	1.000	Ea.	1.143	238.00	31.00	269.00
Painted, 2'-6'' x 6'-8''	1.000	Ea.	1.067	250.00	29.00	279.00
2'-8'' x 6'-8''	1.000	Ea.	1.143	250.00	31.00	281.00
3'-0'' x 6'-8''	1.000	Ea.	1.143	262.00	31.00	293.00
Wood, combination, storm & screen, crossbuck, 2'-6'' x 6'-9''	1.000	Ea.	1.455	227.00	39.50	266.50
2'-8'' x 6'-9''	1.000	Ea.	1.600	227.00	43.00	270.00
3'-0'' x 6'-9''	1.000	Ea.	1.778	232.00	48.00	280.00
Full lite, 2'-6'' x 6'-9''	1.000	Ea.	1.455	221.00	39.50	260.50
2'-8'' x 6'-9''	1.000	Ea.	1.600	221.00	43.00	264.00
3'-0'' x 6'-9''	1.000	Ea.	1.778	229.00	48.00	277.00
Windows						
Windows, aluminum, combination storm & screen, basement, 1'-10'' x 1'-0''	1.000	Ea.	.533	18.35	14.35	32.70
2'-9'' x 1'-6''	1.000	Ea.	.533	24.50	14.35	38.85
3'-4'' x 2'-0''	1.000	Ea.	.533	29.50	14.35	43.85
Double hung, anodized, 2'-0'' x 3'-5''	1.000	Ea.	.533	86.50	14.35	100.85
2'-6'' x 5'-0''	1.000	Ea.	.571	131.00	15.40	146.40
4'-0'' x 6'-0''	1.000	Ea.	.640	235.00	17.20	252.20
Painted, 2'-0'' x 3'-5''	1.000	Ea.	.533	81.50	14.35	95.85
2'-6'' x 5'-0''	1.000	Ea.	.571	131.00	15.40	146.40
4'-0'' x 6'-0''	1.000	Ea.	.640	235.00	17.20	252.20
Fixed window, anodized, 4'-6'' x 4'-6''	1.000	Ea.	.640	102.00	17.20	119.20
5'-8'' x 4'-6''	1.000	Ea.	.800	105.00	21.50	126.50
Painted, 4'-6'' x 4'-6''	1.000	Ea.	.640	90.00	17.20	107.20
5'-8'' x 4'-6''	1.000	Ea.	.800	95.00	21.50	116.50

Shutters/Blinds Systems

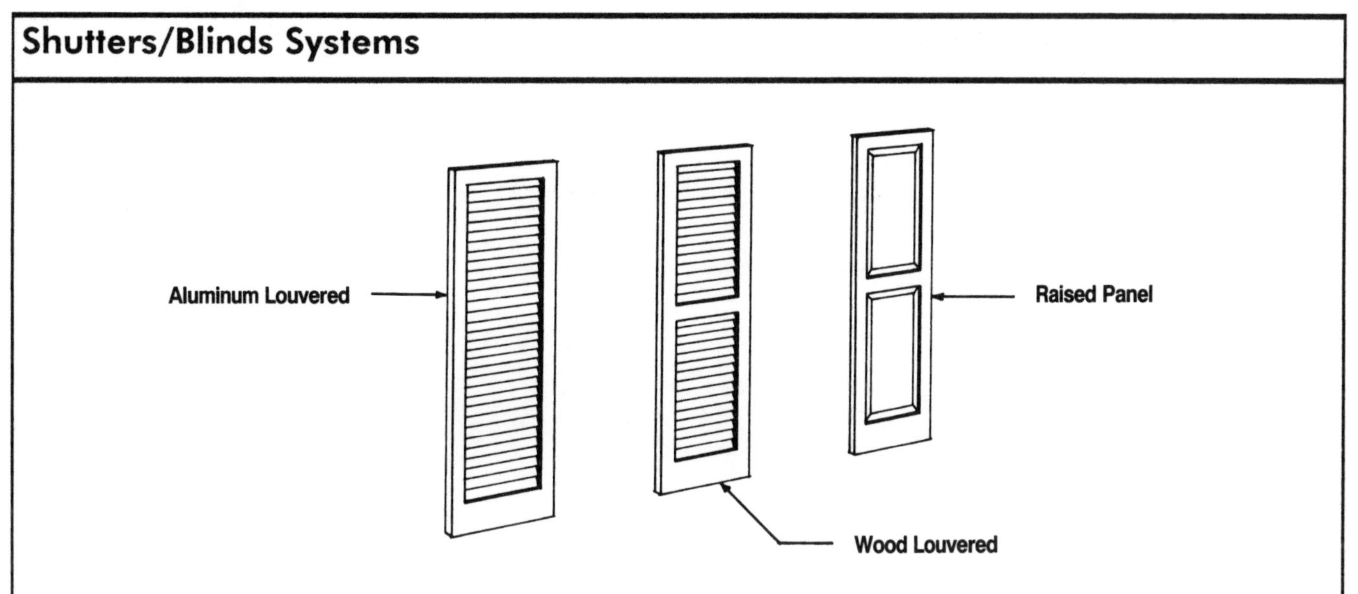

COMPONENT DESCRIPTION	QUAN.	UNIT	LABOR-HOURS	COST EACH		
				MAT.	INST.	TOTAL
Shutters, exterior blinds						
Shutters, exterior blinds, aluminum, louvered, 1'-4'' wide, 3''-0'' long	1.000	Set	.800	38.50	23.50	62.00
4'-0'' long	1.000	Set	.800	42.00	23.50	65.50
5'-4'' long	1.000	Set	.800	45.50	23.50	69.00
6'-8'' long	1.000	Set	.889	64.50	26.00	90.50
Wood, louvered, 1'-2'' wide, 3'-3'' long	1.000	Set	.800	52.00	23.50	75.50
4'-7'' long	1.000	Set	.800	75.00	23.50	98.50
5'-3'' long	1.000	Set	.800	68.00	23.50	91.50
1'-6'' wide, 3'-3'' long	1.000	Set	.800	64.50	23.50	88.00
4'-7'' long	1.000	Set	.800	84.00	23.50	107.50
Polystyrene, solid raised panel, 3'-3'' wide, 3'-0'' long	1.000	Set	.800	80.50	23.50	104.00
3'-11'' long	1.000	Set	.800	90.50	23.50	114.00
5'-3'' long	1.000	Set	.800	112.00	23.50	135.50
6'-8'' long	1.000	Set	.889	150.00	26.00	176.00
Polystyrene, louvered, 1'-2'' wide, 3'-3'' long	1.000	Set	.800	34.00	23.50	57.50
4'-7'' long	1.000	Set	.800	42.00	23.50	65.50
5'-3'' long	1.000	Set	.800	47.50	23.50	71.00
6'-8'' long	1.000	Set	.889	65.50	26.00	91.50
Vinyl, louvered, 1'-2'' wide, 4'-7'' long	1.000	Set	.720	37.00	21.00	58.00
1'-4'' x 6'-8'' long	1.000	Set	.889	64.50	26.00	90.50

Roofing Systems

Gable End Roofing Systems

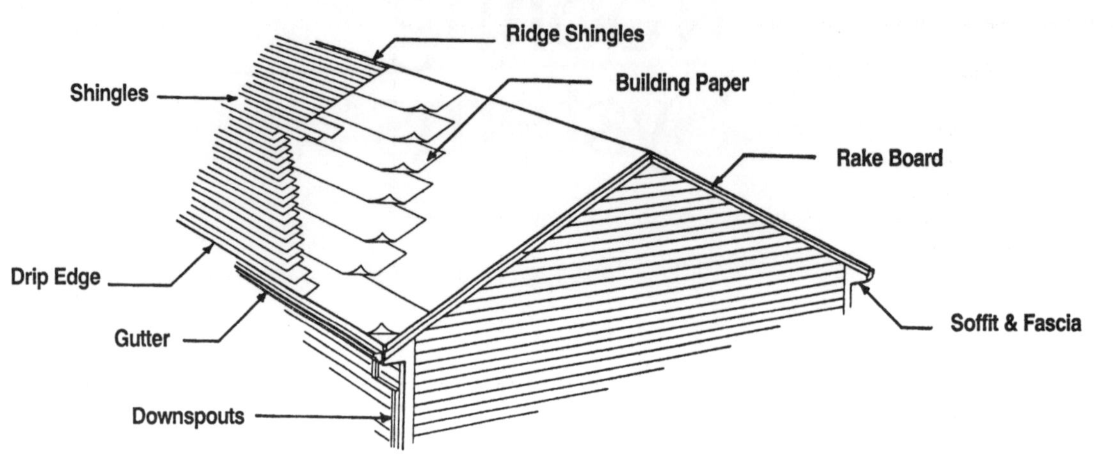

System Description	QUAN.	UNIT	LABOR-HOURS	COST PER S.F.			S.F. EXTENSIONS
				MAT.	INST.	TOTAL	
ASPHALT, ROOF SHINGLES, CLASS A							
Shingles, inorganic class A, 210-235 lb./sq., 4/12 pitch	1.160	S.F.	.014	.30	.45	.75	
Drip edge, metal, 5'' wide	.150	L.F.	.003	.04	.09	.13	
Building paper, #15 felt	1.300	S.F.	.002	.04	.05	.09	
Ridge shingles, asphalt	.042	L.F.	.001	.02	.03	.05	
Soffit & fascia, white painted aluminum, 1' overhang	.083	L.F.	.012	.17	.32	.49	
Rake trim, painted, 1'' x 6''	.040	L.F.	.004	.03	.10	.13	
Gutter, seamless, aluminum painted	.083	L.F.	.006	.09	.17	.26	
Downspouts, aluminum painted	.035	L.F.	.002	.04	.05	.09	
TOTAL			.044	.73	1.26	1.99	
ASPHALT, ROOF SHINGLES, PREMIUM LAMINATED							
Shingles, multi-layered, 260-300 lb./sq., 4/12 pitch	1.160	S.F.	.027	.77	.77	1.54	
Drip edge, metal, 5'' wide	.150	L.F.	.003	.04	.09	.13	
Building paper, #15 felt	1.300	S.F.	.002	.04	.05	.09	
Ridge shingles, asphalt	.042	L.F.	.001	.02	.03	.05	
Rake trim, painted, 1'' x 6''	.040	L.F.	.004	.03	.10	.13	
Soffit & fascia, white painted aluminum, 1' overhang	.083	L.F.	.012	.17	.32	.49	
Gutter, seamless, aluminum painted	.083	L.F.	.006	.09	.17	.26	
Downspouts, aluminum painted	.035	L.F.	.002	.04	.05	.09	
TOTAL			.057	1.20	1.58	2.78	
CLAY TILE, SPANISH TILE, RED							
Clay tile, Spanish tile, red, 4/12 pitch	1.160	S.F.	.053	5.10	1.50	6.60	
Drip edge, metal, 5'' wide	.150	L.F.	.003	.04	.09	.13	
Building paper, #15 felt	1.300	S.F.	.002	.04	.05	.09	
Ridge tiles, clay	.042	L.F.	.002	.38	.05	.43	
Soffit & fascia, white painted aluminum, 1' overhang	.083	L.F.	.012	.17	.32	.49	
Rake trim, painted, 1'' x 6''	.040	L.F.	.004	.03	.10	.13	
Gutter, seamless, aluminum painted	.083	L.F.	.006	.09	.17	.26	
Downspouts, aluminum painted	.035	L.F.	.002	.04	.05	.09	
TOTAL			.084	5.89	2.33	8.22	

Gable End Roofing Systems

System Description	QUAN.	UNIT	LABOR-HOURS	COST PER S.F.			S.F. EXTENSIONS
				MAT.	INST.	TOTAL	
WOOD, CEDAR SHINGLES NO. 1 PERFECTIONS, 18″ LONG							
Shingles, wood, cedar, No. 1 perfections, 4/12 pitch	1.160	S.F.	.035	2.38	1.03	3.41	
Drip edge, metal, 5″ wide	.150	L.F.	.003	.04	.09	.13	
Building paper, #15 felt	1.300	S.F.	.002	.04	.05	.09	
Ridge shingles, cedar	.042	L.F.	.001	.08	.04	.12	
Soffit & fascia, white painted aluminum, 1′ overhang	.083	L.F.	.012	.17	.32	.49	
Rake trim, painted, 1″ x 6″	.040	L.F.	.004	.03	.10	.13	
Gutter, seamless, aluminum, painted	.083	L.F.	.006	.09	.17	.26	
Downspouts, aluminum, painted	.035	L.F.	.002	.04	.05	.09	
TOTAL			.065	2.87	1.85	4.72	
WOOD, CEDAR SHINGLES, RESQUARED AND REBUTTED, 18″ LONG							
Shingles, Wood, cedar, resquared & rebutted, 4/12 pitch	1.160	S.F.	.032	1.32	.94	2.26	
Drip edge, metal, 5″ wide	.150	L.F.	.003	.04	.09	.13	
Building paper, #15 felt	1.300	S.F.	.002	.04	.05	.09	
Ridge shingles, cedar	.042	L.F.	.001	.08	.04	.12	
Soffit & fascia, white painted aluminum, 1′ overhang	.083	L.F.	.012	.17	.32	.49	
Rake trim, painted, 1″ x 6″	.040	L.F.	.004	.03	.10	.13	
Gutter, seamless, aluminum, painted	.083	L.F.	.006	.09	.17	.26	
Downspouts, aluminum, painted	.035	L.F.	.002	.04	.05	.09	
TOTAL			.062	1.81	1.76	3.57	
WOOD SHAKES							
Shingles, hand split, 24″ long, 10″ exposure 4/12 pitch	1.160	S.F.	.038	1.98	1.13	3.11	
Drip edge, metal, 5″ wide	.150	L.F.	.003	.04	.09	.13	
Building paper, #15 felt	1.300	S.F.	.002	.04	.05	.09	
Ridge shingles, shakes	.042	L.F.	.001	.08	.04	.12	
Soffit & fascia, white painted aluminum, 1′ overhang	.083	L.F.	.012	.17	.32	.49	
Rake trim, painted, 1″ x 6″	.040	L.F.	.004	.03	.10	.13	
Gutter, seamless, aluminum, painted	.083	L.F.	.006	.09	.17	.26	
Downspouts, aluminum, painted	.035	L.F.	.002	.04	.05	.09	
TOTAL			.068	2.47	1.95	4.42	

The prices in these systems are based on a square foot of plan area.
All quantities have been adjusted accordingly.

Gable End Roofing System Components

Component Description	QUAN.	UNIT	LABOR-HOURS	COST PER S.F.		
				MAT.	INST.	TOTAL
Shingles						
Shingles, asphalt, inorganic, class A, 210-235 lb./sq., 4/12 pitch	1.160	S.F.	.014	.30	.45	.75
8/12 pitch	1.330	S.F.	.015	.33	.50	.83
Mission tile, red, 4/12 pitch	1.160	S.F.	.083	8.05	2.35	10.40
8/12 pitch	1.330	S.F.	.090	8.70	2.55	11.25
Slate, Buckingham, Virginia, black, 4/12 pitch	1.160	S.F.	.055	6.70	1.55	8.25
8/12 pitch	1.330	S.F.	.059	7.30	1.68	8.98
Wood, No. 1 red cedar, 5X, 16″ long, 5″ exposure, 4/12 pitch	1.160	S.F.	.038	2.05	1.13	3.18
8/12 pitch	1.330	S.F.	.042	2.22	1.22	3.44
Fire retardant, 4/12 pitch	1.160	S.F.	.038	2.52	1.13	3.65
8/12 pitch	1.330	S.F.	.042	2.73	1.22	3.95
Wood shakes hand split, 24″ long, 10″ exposure, 4/12 pitch	1.160	S.F.	.038	1.98	1.13	3.11
18″ long, 8″ exposure, 4/12 pitch	1.160	S.F.	.048	1.81	1.40	3.21

Hip Roof Roofing Systems

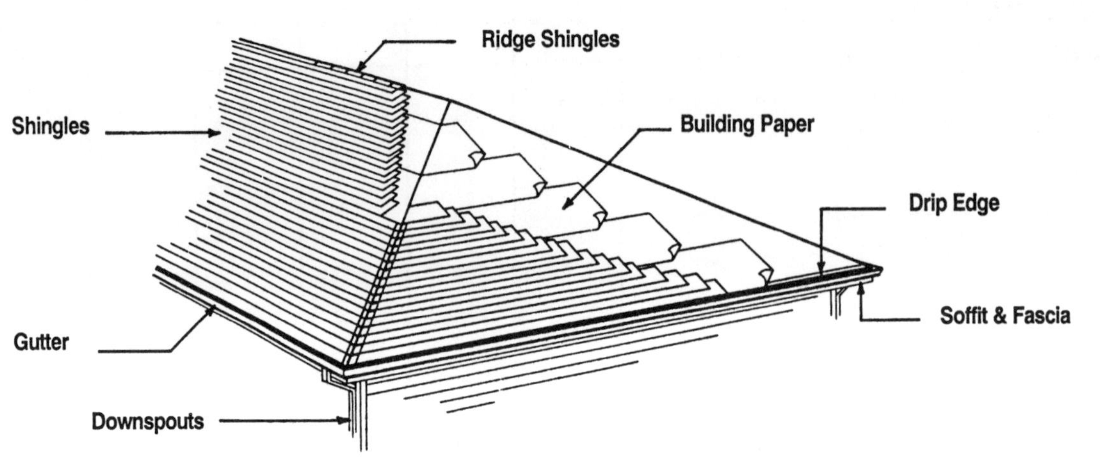

System Description	QUAN.	UNIT	LABOR-HOURS	COST PER S.F.			S.F. EXTENSIONS
				MAT.	INST.	TOTAL	
ASPHALT, ROOF SHINGLES, CLASS A							
Shingles, inorganic, class A, 210-235 lb./sq. 4/12 pitch	1.570	S.F.	.018	.40	.61	1.01	
Drip edge, metal, 5″ wide	.122	L.F.	.002	.03	.07	.10	
Building paper, #15 asphalt felt	1.800	S.F.	.002	.05	.07	.12	
Ridge shingles, asphalt	.075	L.F.	.002	.04	.05	.09	
Soffit & fascia, white painted aluminum, 1′ overhang	.120	L.F.	.017	.25	.47	.72	
Gutter, seamless, aluminum, painted	.120	L.F.	.008	.13	.25	.38	
Downspouts, aluminum, painted	.035	L.F.	.002	.04	.05	.09	
TOTAL			.051	.94	1.57	2.51	
ASPHALT, ROOF SHINGLES, LAMINATED, MULTI-LAYERED							
Laminated, multi-layered, 240-260 lb./sq., 4/12 pitch	1.570	S.F.	.028	.73	.80	1.53	
Drip edge, metal, 5″ wide	.122	L.F.	.002	.03	.07	.10	
Building paper, #15 asphalt felt	1.800	S.F.	.002	.05	.07	.12	
Ridge shingles, asphalt	.075	L.F.	.002	.04	.05	.09	
Soffit & fascia, white painted aluminum, 1′ overhang	.120	L.F.	.017	.25	.47	.72	
Gutter, seamless, aluminum, painted	.120	L.F.	.008	.13	.25	.38	
Downspouts, aluminum, painted	.035	L.F.	.002	.04	.05	.09	
TOTAL			.061	1.27	1.76	3.03	
MISSION TILE, RED							
Mission tile, red, 4/12 pitch	1.570	S.F.	.111	10.72	3.14	13.86	
Drip edge, metal, 5″ wide	.122	L.F.	.002	.03	.07	.10	
Building paper, #15 asphalt felt	1.800	S.F.	.002	.05	.07	.12	
Ridge tiles, clay	.075	L.F.	.003	.69	.08	.77	
Soffit & fascia, white painted aluminum, 1′ overhang	.120	L.F.	.017	.25	.47	.72	
Gutter, seamless, aluminum, painted	.120	L.F.	.008	.13	.25	.38	
Downspouts, aluminum, painted	.035	L.F.	.002	.04	.05	.09	
TOTAL			.145	11.91	4.13	16.04	

Hip Roof Roofing Systems

System Description	QUAN.	UNIT	LABOR-HOURS	COST PER S.F.			S.F. EXTENSIONS
				MAT.	INST.	TOTAL	
WOOD, CEDAR SHINGLES, NO. 1 PERFECTIONS, 18″ LONG							
Shingles, red cedar, No. 1 perfections, 5″ exp., 4/12 pitch	1.570	S.F.	.047	3.17	1.37	4.54	
Drip edge, metal, 5″ wide	.122	L.F.	.002	.03	.07	.10	
Building paper, #15 asphalt felt	1.800	S.F.	.002	.05	.07	.12	
Ridge shingles, wood, cedar	.075	L.F.	.002	.14	.06	.20	
Soffit & fascia, white painted aluminum, 1′ overhang	.120	L.F.	.017	.25	.47	.72	
Gutter, seamless, aluminum, painted	.120	L.F.	.008	.13	.25	.38	
Downspouts, aluminum, painted	.035	L.F.	.002	.04	.05	.09	
TOTAL			.080	3.81	2.34	6.15	
WOOD, FIRE RETARDANT, CEDAR SHINGLES, NO. 1, 18″ LONG							
Shingles, red cedar, fire retardant, 5″ exp., 4/12 pitch	1.570	S.F.	.047	3.85	1.37	5.22	
Drip edge, metal, 5″ wide	.122	L.F.	.002	.03	.07	.10	
Building paper, #15 asphalt felt	1.800	S.F.	.002	.05	.07	.12	
Ridge shingles, wood, cedar	.075	L.F.	.002	.14	.06	.20	
Soffit & fascia, white painted aluminum, 1′ overhang	.120	L.F.	.017	.25	.47	.72	
Gutter, seamless, aluminum, painted	.120	L.F.	.008	.13	.25	.38	
Downspouts, aluminum, painted	.035	L.F.	.002	.04	.05	.09	
TOTAL			.080	4.49	2.34	6.83	
WOOD SHAKES							
Shingles, hand split, 24″ long, 10″ exp., 4/12 pitch	1.510	S.F.	.051	2.64	1.50	4.14	
Drip edge, metal, 5″ wide	.122	L.F.	.002	.03	.07	.10	
Building paper, #15 asphalt felt	1.800	S.F.	.002	.05	.07	.12	
Ridge shingles, shakes	.075	L.F.	.002	.14	.06	.20	
Soffit & fascia, white painted aluminum, 1′ overhang	.120	L.F.	.017	.25	.47	.72	
Gutter, seamless, aluminum, painted	.012	L.F.	.008	.13	.25	.38	
Downspouts, aluminum, painted	.035	L.F.	.002	.04	.05	.09	
TOTAL			.084	3.28	2.47	5.75	

The prices in these systems are based on a square foot of plan area.
All quantities have been adjusted accordingly.

Hip Roof Roofing System Components

Component Description	QUAN.	UNIT	LABOR-HOURS	COST PER S.F.		
				MAT.	INST.	TOTAL
Shingles						
Shingles, asphalt, inorganic, class A, 210-235 lb./sq., 4/12 pitch	1.570	S.F.	.018	.40	.61	1.01
8/12 pitch	1.850	S.F.	.022	.48	.72	1.20
Laminated, multi-layered, 240-260 lb./sq., 4/12 pitch	1.570	S.F.	.028	.73	.80	1.53
8/12 pitch	1.850	S.F.	.034	.86	.95	1.81
Prem. laminated, multi-layered, 260-300 lb./sq., 4/12 pitch	1.570	S.F.	.037	1.03	1.03	2.06
8/12 pitch	1.850	S.F.	.043	1.23	1.23	2.46
Clay tile, Spanish tile, red, 4/12 pitch	1.570	S.F.	.071	6.80	2.00	8.80
8/12 pitch	1.850	S.F.	.084	8.10	2.38	10.48
Slate, Buckingham, Virginia, black, 4/12 pitch	1.570	S.F.	.073	8.95	2.06	11.01
8/12 pitch	1.850	S.F.	.087	10.65	2.45	13.10
Vermont, black or grey, 4/12 pitch	1.570	S.F.	.073	6.70	2.06	8.76
8/12 pitch	1.850	S.F.	.087	8.00	2.45	10.45
Wood, red cedar, No.1 5X, 16″ long, 5″ exposure, 4/12 pitch	1.570	S.F.	.051	2.74	1.50	4.24
8/12 pitch	1.850	S.F.	.061	3.25	1.79	5.04
Fire retardant, 4/12 pitch	1.570	S.F.	.051	3.37	1.50	4.87
8/12 pitch	1.850	S.F.	.061	4.00	1.79	5.79

Gambrel Roofing Systems

System Description	QUAN.	UNIT	LABOR-HOURS	COST PER S.F.			S.F. EXTENSIONS
				MAT.	INST.	TOTAL	
ASPHALT, ROOF SHINGLES, CLASS A							
Shingles, asphalt, inorganic, class A, 210-235 lb./sq.	1.450	S.F.	.017	.38	.57	.95	
Drip edge, metal, 5'' wide	.146	L.F.	.003	.04	.09	.13	
Building paper, #15 asphalt felt	1.500	S.F.	.002	.04	.06	.10	
Ridge shingles, asphalt	.042	L.F.	.001	.02	.03	.05	
Soffit & fascia, painted aluminum, 1' overhang	.083	L.F.	.012	.17	.32	.49	
Rake, trim, painted, 1'' x 6''	.063	L.F.	.006	.05	.14	.19	
Gutter, seamless, alumunum, painted	.083	L.F.	.006	.09	.17	.26	
Downspouts, aluminum, painted	.042	L.F.	.002	.05	.06	.11	
TOTAL			.049	.84	1.44	2.28	
ASPHALT, ROOF SHINGLES, LAMINATED, MULTI-LAYERED							
Laminated, multi-layered, 240-260 lb./sq.	1.450	S.F.	.027	.68	.75	1.43	
Drip edge, metal, 5'' wide	.146	L.F.	.003	.04	.09	.13	
Building paper, #15 asphalt felt	1.500	S.F.	.002	.04	.06	.10	
Ridge shingles, asphalt	.042	L.F.	.001	.02	.03	.05	
Soffit & fascia, painted aluminum, 1' overhang	.083	L.F.	.012	.17	.32	.49	
Rake, trim, painted, 1'' x 6''	.063	L.F.	.006	.05	.14	.19	
Gutter, seamless, alumunum, painted	.083	L.F.	.006	.09	.17	.26	
Downspouts, aluminum, painted	.042	L.F.	.002	.05	.06	.11	
TOTAL			.059	1.14	1.62	2.76	
ASPHALT, ROOF SHINGLES, PREMIUM LAMINATED							
Premium laminated, multi-layered, 260-300 lb./sq.	1.450	S.F.	.034	.97	.97	1.94	
Drip edge, metal, 5'' wide	.146	L.F.	.003	.04	.09	.13	
Building paper, #15 asphalt felt	1.500	S.F.	.002	.04	.06	.10	
Ridge shingles, asphalt	.042	L.F.	.001	.02	.03	.05	
Soffit & fascia, painted aluminum, 1' overhang	.083	L.F.	.012	.17	.32	.49	
Rake, trim, painted, 1'' x 6''	.063	L.F.	.006	.05	.14	.19	
Gutter, seamless, alumunum, painted	.083	L.F.	.006	.09	.17	.26	
Downspouts, aluminum, painted	.042	L.F.	.002	.05	.06	.11	
TOTAL			.066	1.43	1.84	3.27	

Gambrel Roofing Systems

System Description	QUAN.	UNIT	LABOR-HOURS	COST PER S.F.			S.F. EXTENSIONS
				MAT.	INST.	TOTAL	
VIRGINIA SLATE							
Slate, Buckingham, Virginia, black	1.450	S.F.	.069	8.40	1.94	10.34	
Drip edge, metal, 5" wide	.146	L.F.	.003	.04	.09	.13	
Building paper, #15 asphalt felt	1.500	S.F.	.002	.04	.06	.10	
Ridge slate	.042	L.F.	.002	.20	.05	.25	
Soffit & fascia, painted aluminum, 1' overhang	.083	L.F.	.012	.17	.32	.49	
Rake, trim, painted, 1" x 6"	.063	L.F.	.006	.05	.14	.19	
Gutter, seamless, alumunum, painted	.083	L.F.	.006	.09	.17	.26	
Downspouts, aluminum, painted	.042	L.F.	.002	.05	.06	.11	
TOTAL			.102	9.04	2.83	11.87	
WOOD, CEDAR SHINGLES, NO. 1 PERFECTIONS, 18" LONG							
Shingles, wood, red cedar, No. 1 perfections, 5" exposure	1.450	S.F.	.044	2.97	1.28	4.25	
Drip edge, metal, 5" wide	.146	L.F.	.003	.04	.09	.13	
Building paper, #15 asphalt felt	1.500	S.F.	.002	.04	.06	.10	
Ridge shingles, wood	.042	L.F.	.001	.08	.04	.12	
Soffit & fascia, white painted aluminum, 1' overhang	.083	L.F.	.012	.17	.32	.49	
Rake, trim, painted, 1" x 6"	.063	L.F.	.004	.05	.09	.14	
Gutter, seamless, aluminum, painted	.083	L.F.	.006	.09	.17	.26	
Downspouts, aluminum, painted	.042	L.F.	.002	.05	.06	.11	
TOTAL			.074	3.49	2.11	5.60	
WOOD SHAKES							
Shingles, hand split, 24" long, 10" exp., plain	1.450	S.F.	.048	2.48	1.41	3.89	
Drip edge, metal, 5" wide	.146	L.F.	.003	.04	.09	.13	
Building paper, #15 felt	1.500	S.F.	.002	.04	.06	.10	
Ridge shingles, wood, shakes	.042	S.F.	.001	.08	.04	.12	
Soffit & fascia, white painted aluminum, 1' overhang	.083	L.F.	.012	.17	.32	.49	
Rake, trim, painted, 1" x 6"	.063	L.F.	.004	.05	.09	.14	
Gutter, seamless, aluminum, painted	.083	L.F.	.006	.09	.17	.26	
Downspouts, aluminum, painted	.042	L.F.	.002	.05	.06	.11	
TOTAL			.078	3.00	2.24	5.24	

The prices in this system are based on a square foot of plan area.
All quantities have been adjusted accordingly.

Gambrel Roofing System Components

Component Description	QUAN.	UNIT	LABOR-HOURS	COST PER S.F.		
				MAT.	INST.	TOTAL
Shingles						
Wood, red cedar, No.1 5X, 16" long, 5" exposure, plain	1.450	S.F.	.048	2.57	1.41	3.98
Fire retardant	1.450	S.F.	.048	3.16	1.41	4.57
Shakes, hand split, 24" long, 10" exposure, plain	1.450	S.F.	.048	2.48	1.41	3.89
18" long, 8" exposure, plain	1.450	S.F.	.060	2.27	1.76	4.03
Soffit & fascia						
Soffit & fascia, aluminum, vented, 1' overhang	.083	L.F.	.012	.17	.32	.49
2' overhang	.083	L.F.	.013	.25	.36	.61
Wood board fascia, plywood soffit, 1' overhang	.083	L.F.	.004	.01	.09	.10
2' overhang	.083	L.F.	.006	.02	.14	.16

Mansard Roofing Systems

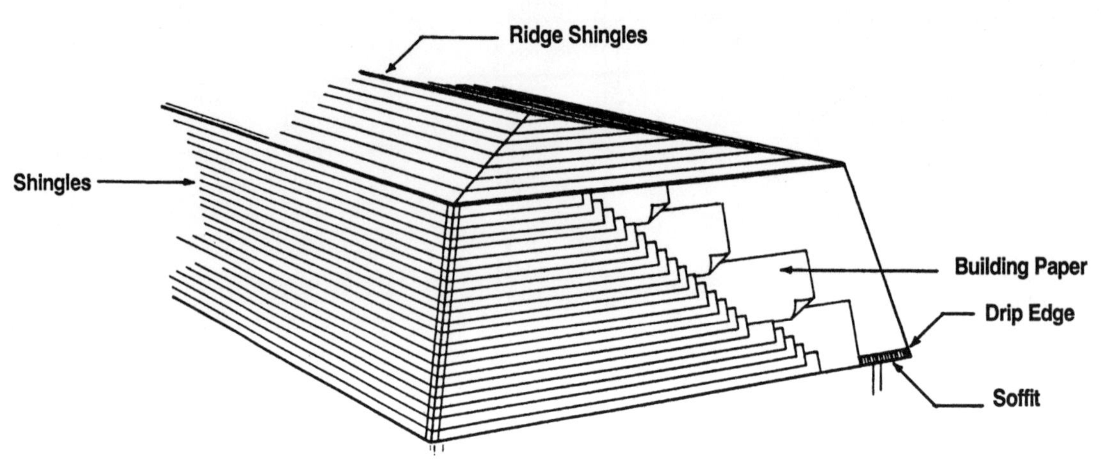

System Description	QUAN.	UNIT	LABOR-HOURS	COST PER S.F.			S.F. EXTENSIONS
				MAT.	INST.	TOTAL	
ASPHALT, ROOF SHINGLES, CLASS A							
Shingles, standard inorganic class A 210-235 lb./sq.	2.210	S.F.	.025	.55	.83	1.38	
Drip edge, metal, 5″ wide	.122	L.F.	.002	.03	.07	.10	
Building paper, #15 asphalt felt	2.300	S.F.	.003	.06	.09	.15	
Ridge shingles, asphalt	.090	L.F.	.002	.05	.06	.11	
Soffit & fascia, white painted aluminum, 1' overhang	.122	L.F.	.018	.25	.48	.73	
Gutter, seamless, aluminum, painted	.122	L.F.	.008	.13	.25	.38	
Downspouts, aluminum, painted	.042	L.F.	.002	.05	.06	.11	
TOTAL			.060	1.12	1.84	2.96	
ASPHALT, ROOF SHINGLES, PREMIUM LAMINATED							
Premium laminated, multi-layered, 260-300 lb./sq.	2.210	S.F.	.050	1.42	1.42	2.84	
Drip edge, metal, 5″ wide	.122	L.F.	.002	.03	.07	.10	
Building paper, #15 asphalt felt	2.300	S.F.	.003	.06	.09	.15	
Ridge shingles, asphalt	.090	L.F.	.002	.05	.06	.11	
Soffit & fascia, white painted aluminum, 1' overhang	.122	L.F.	.018	.25	.48	.73	
Gutter, seamless, aluminum, painted	.122	L.F.	.008	.13	.25	.38	
Downspouts, aluminum, painted	.042	L.F.	.002	.05	.06	.11	
TOTAL			.085	1.99	2.43	4.42	
VERMONT SLATE, BLACK OR GREY							
Vermont, black or grey	2.210	S.F.	.101	9.24	2.84	12.08	
Drip edge, metal, 5″ wide	.122	L.F.	.002	.03	.07	.10	
Building paper, #15 asphalt felt	2.300	S.F.	.003	.06	.09	.15	
Ridge slate	.090	L.F.	.004	.43	.10	.53	
Soffit & fascia, white painted aluminum, 1' overhang	.122	L.F.	.018	.25	.48	.73	
Gutter, seamless, aluminum, painted	.122	L.F.	.008	.13	.25	.38	
Downspouts, aluminum, painted	.042	L.F.	.002	.05	.06	.11	
TOTAL			.138	10.19	3.89	14.08	

Mansard Roofing Systems

System Description	QUAN.	UNIT	LABOR-HOURS	COST PER S.F.			S.F. EXTENSIONS
				MAT.	INST.	TOTAL	
WOOD, CEDAR SHINGLES, NO. 1 PERFECTIONS, 18" LONG							
Shingles, wood, red cedar, No. 1 perfections, 5" exposure	2.210	S.F.	.064	4.36	1.88	6.24	
Drip edge, metal, 5" wide	.122	L.F.	.002	.03	.07	.10	
Building paper, #15 asphalt felt	2.300	S.F.	.003	.06	.09	.15	
Ridge shingles, wood	.090	L.F.	.003	.16	.08	.24	
Soffit & fascia, white painted aluminum, 1' overhang	.122	L.F.	.018	.25	.48	.73	
Gutter, seamless, aluminum, painted	.122	L.F.	.008	.13	.25	.38	
Downspouts, aluminum, painted	.042	L.F.	.002	.05	.06	.11	
TOTAL			.100	5.04	2.91	7.95	
WOOD, CEDAR SHINGLES, FIRE RETARD, NO. 1 PERFECTIONS							
Shingles, wood, red cedar, fire retardant, 5" exp.	2.210	S.F.	.070	4.63	2.07	6.70	
Drip edge, metal, 5" wide	.122	L.F.	.002	.03	.07	.10	
Building paper, #15 asphalt felt	2.300	S.F.	.003	.06	.09	.15	
Ridge shingles, wood	.090	L.F.	.003	.16	.08	.24	
Soffit & fascia, white painted aluminum, 1' overhang	.122	L.F.	.018	.25	.48	.73	
Gutter, seamless, aluminum, painted	.122	L.F.	.008	.13	.25	.38	
Downspouts, aluminum, painted	.042	L.F.	.002	.05	.06	.11	
TOTAL			.106	5.31	3.10	8.41	
WOOD SHAKES							
Shingles, hand split, 24" long, 10" exposure, plain	2.210	S.F.	.070	3.63	2.07	5.70	
Drip edge, metal, 5" wide	.122	L.F.	.002	.03	.07	.10	
Building paper, #15 asphalt felt	2.300	S.F.	.003	.06	.09	.15	
Ridge shingles, shakes	.090	L.F.	.003	.16	.08	.24	
Soffit & fascia, white painted aluminum, 1' overhang	.122	L.F.	.018	.25	.48	.73	
Gutter, seamless, aluminum, painted	.122	L.F.	.008	.13	.25	.38	
Downspouts, aluminum, painted	.042	L.F.	.002	.05	.06	.11	
TOTAL			.106	4.31	3.10	7.41	

The prices in these systems are based on a square foot of plan area.
All quantities have been adjusted accordingly.

Mansard Roofing System Components

Component Description	QUAN.	UNIT	LABOR-HOURS	COST PER S.F.		
				MAT.	INST.	TOTAL
Shingles						
Laminated, multi-layered, 240-260 lb./sq.	2.210	S.F.	.039	1.00	1.10	2.10
Slate Buckingham, Virginia, black	2.210	S.F.	.101	12.30	2.84	15.14
Wood, red cedar, No.1 5X, 16" long, 5" exposure, plain	2.210	S.F.	.070	3.76	2.07	5.83
Fire retardant	2.210	S.F.	.070	4.63	2.07	6.70
Resquared & rebutted, 18" long, 6" exposure, plain	2.210	S.F.	.059	2.42	1.73	4.15
Fire retardant	2.210	S.F.	.059	3.36	1.73	5.09
Shakes, hand split, 18" long, 8" exposure, plain	2.210	S.F.	.088	3.32	2.57	5.89
Fire retardant	2.210	S.F.	.088	4.19	2.57	6.76
Downspout						
Downspout 2" x 3", aluminum, one story house	.042	L.F.	.002	.04	.06	.10
Two story house	.070	L.F.	.003	.06	.09	.15
Vinyl, one story house	.042	L.F.	.002	.04	.06	.10
Two story house	.070	L.F.	.003	.06	.09	.15

Shed Roofing Systems

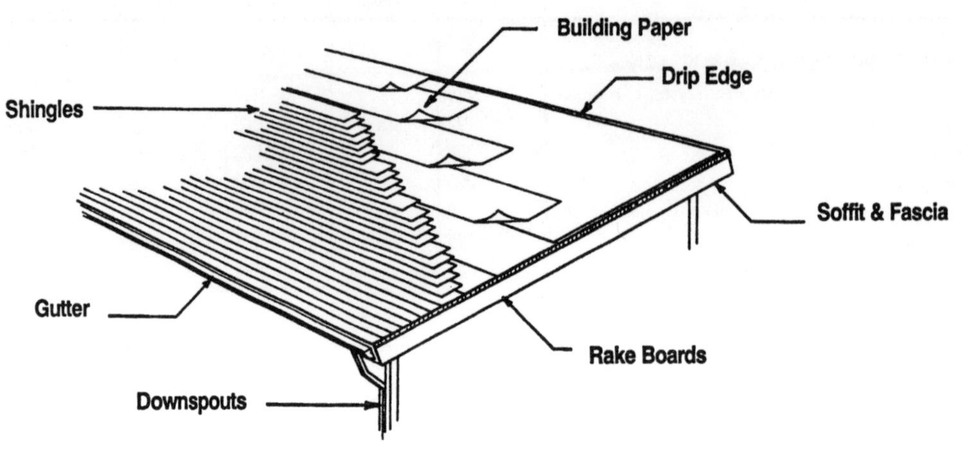

System Description	QUAN.	UNIT	LABOR-HOURS	COST PER S.F.			S.F. EXTENSIONS
				MAT.	INST.	TOTAL	
ASPHALT, ROOF SHINGLES, CLASS A							
Shingles, inorganic class A 210-235 lb./sq. 4/12 pitch	1.230	S.F.	.015	.33	.50	.83	
Drip edge, metal, 5" wide	.100	L.F.	.002	.03	.06	.09	
Building paper, #15 asphalt felt	1.300	S.F.	.002	.04	.05	.09	
Soffit & fascia, white painted aluminum, 1' overhang	.080	L.F.	.012	.17	.31	.48	
Rake trim, painted, 1" x 6"	.043	L.F.	.004	.03	.10	.13	
Gutter, seamless, aluminum, painted	.040	L.F.	.003	.04	.08	.12	
Downspouts, painted aluminum	.020	L.F.	.001	.02	.03	.05	
TOTAL			.039	.66	1.13	1.79	
ASPHALT, ROOF SHINGLES, PREMIUM LAMINATED							
Shingles, premium laminated, 210-235 lb./sq. 4/12 pitch	1.230	S.F.	.027	.77	.77	1.54	
Drip edge, metal, 5" wide	.100	L.F.	.002	.03	.06	.09	
Building paper, #15 asphalt felt	1.300	S.F.	.002	.04	.05	.09	
Soffit & fascia, white painted aluminum, 1' overhang	.080	L.F.	.012	.17	.31	.48	
Rake trim, painted, 1" x 6"	.043	L.F.	.004	.03	.10	.13	
Gutter, seamless, aluminum, painted	.040	L.F.	.003	.04	.08	.12	
Downspouts, painted aluminum	.020	L.F.	.001	.02	.03	.05	
TOTAL			.051	1.10	1.40	2.50	
CLAY TILE, SPANISH TILE							
Clay tile, Spanish tile, red, 4/12 pitch	1.230	S.F.	.053	5.10	1.50	6.60	
Drip edge, metal, 5" wide	.100	L.F.	.002	.03	.06	.09	
Building paper, #15 asphalt felt	1.300	S.F.	.002	.04	.05	.09	
Soffit & fascia, white painted aluminum, 1' overhang	.080	L.F.	.012	.17	.31	.48	
Rake trim, painted, 1" x 6"	.043	L.F.	.004	.03	.10	.13	
Gutter, seamless, aluminum, painted	.040	L.F.	.003	.04	.08	.12	
Downspouts, painted aluminum	.020	L.F.	.001	.02	.03	.05	
TOTAL			.077	5.43	2.13	7.56	

Shed Roofing Systems

System Description	QUAN.	UNIT	LABOR-HOURS	COST PER S.F.			S.F. EXTENSIONS
				MAT.	INST.	TOTAL	
WOOD, CEDAR SHINGLES, NO. 1 PERFECTIONS, 18" LONG							
Shingles, red cedar, No. 1 perfections, 5" exp., 4/12 pitch	1.230	S.F.	.035	2.38	1.03	3.41	
Drip edge, metal, 5" wide	.100	L.F.	.002	.03	.06	.09	
Building paper, #15 asphalt felt	1.300	S.F.	.002	.04	.05	.09	
Soffit & fascia, white painted aluminum, 1' overhang	.080	L.F.	.012	.17	.31	.48	
Rake trim, painted, 1" x 6"	.043	L.F.	.003	.03	.07	.10	
Gutter, seamless, aluminum, painted	.040	L.F.	.003	.04	.08	.12	
Downspouts, painted aluminum	.020	L.F.	.001	.02	.03	.05	
TOTAL			.058	2.71	1.63	4.34	
WOOD, CEDAR SHINGLES, RESQUARED & REBUTTED 18" LONG							
Resquared & rebutted, 18" long, 5" exposure, 4/12 pitch	1.230	S.F.	.032	1.32	.94	2.26	
Drip edge, metal, 5" wide	.100	L.F.	.002	.03	.06	.09	
Building paper, #15 asphalt felt	1.300	S.F.	.002	.04	.05	.09	
Soffit & fascia, white painted aluminum, 1' overhang	.080	L.F.	.012	.17	.31	.48	
Rake trim, painted, 1" x 6"	.043	L.F.	.003	.03	.07	.10	
Gutter, seamless, aluminum, painted	.040	L.F.	.003	.04	.08	.12	
Downspouts, painted aluminum	.020	L.F.	.001	.02	.03	.05	
TOTAL			.055	1.65	1.54	3.19	
WOOD SHAKES							
Shingles, hand split, 24" long, 10" exp., 4/12 pitch	1.230	S.F.	.038	1.98	1.13	3.11	
Drip edge, metal, 5" wide	.100	L.F.	.002	.03	.06	.09	
Building paper, #15 asphalt felt	1.300	S.F.	.002	.04	.05	.09	
Soffit & fascia, white painted aluminum, 1' overhang	.080	L.F.	.012	.17	.31	.48	
Rake trim, painted, 1" x 6"	.043	L.F.	.003	.03	.07	.10	
Gutter, seamless, aluminum, painted	.040	L.F.	.003	.04	.08	.12	
Downspouts, painted aluminum	.020	L.F.	.001	.02	.03	.05	
TOTAL			.061	2.31	1.73	4.04	

The prices in these systems are based on a square foot of plan area.
All quantities have been adjusted accordingly.

Shed Roofing System Components

Component Description	QUAN.	UNIT	LABOR-HOURS	COST PER S.F.		
				MAT.	INST.	TOTAL
Shingles						
Shingles, asphalt, inorganic, class A, 210-235 lb./sq., 4/12 pitch	1.230	S.F.	.014	.30	.45	.75
8/12 pitch	1.330	S.F.	.015	.33	.50	.83
Laminated, multi-layered, 240-260 lb./sq. 4/12 pitch	1.230	S.F.	.021	.55	.60	1.15
8/12 pitch	1.330	S.F.	.023	.59	.65	1.24
Premium laminated, multi-layered, 260-300 lb./sq. 4/12 pitch	1.230	S.F.	.027	.77	.77	1.54
8/12 pitch	1.330	S.F.	.030	.84	.84	1.68
Mission tile, red, 4/12 pitch	1.230	S.F.	.083	8.05	2.35	10.40
8/12 pitch	1.330	S.F.	.090	8.70	2.55	11.25
French tile, red, 4/12 pitch	1.230	S.F.	.071	7.25	2.00	9.25
8/12 pitch	1.330	S.F.	.077	7.85	2.17	10.02
Slate, Buckingham, Virginia, black, 4/12 pitch	1.230	S.F.	.055	6.70	1.55	8.25
8/12 pitch	1.330	S.F.	.059	7.30	1.68	8.98
Vermont, black or grey, 4/12 pitch	1.230	S.F.	.055	5.05	1.55	6.60
8/12 pitch	1.330	S.F.	.059	5.45	1.68	7.13
Wood, red cedar, No.1 5X, 16" long, 5" exposure, 4/12 pitch	1.230	S.F.	.038	2.05	1.13	3.18
8/12 pitch	1.330	S.F.	.042	2.22	1.22	3.44

Gable Dormer Roofing Systems

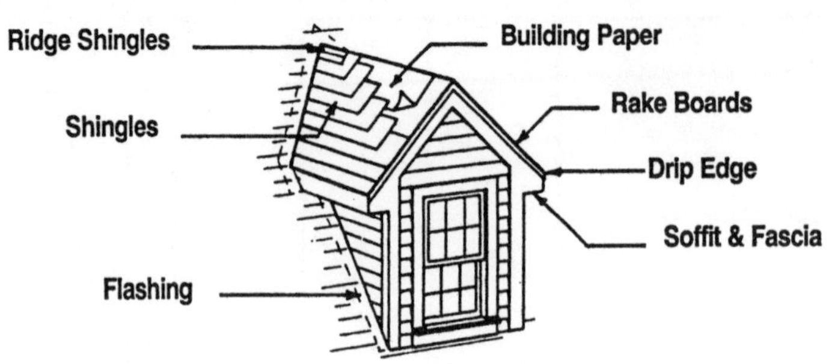

System Description	QUAN.	UNIT	LABOR-HOURS	COST PER S.F.			S.F. EXTENSIONS
				MAT.	INST.	TOTAL	
ASPHALT, ROOF SHINGLES, CLASS A							
Shingles, standard inorganic class A 210-235 lb./sq	1.400	S.F.	.016	.35	.53	.88	
Drip edge, metal, 5" wide	.220	L.F.	.004	.06	.13	.19	
Building paper, #15 asphalt felt	1.500	S.F.	.002	.04	.06	.10	
Ridge shingles, asphalt	.280	L.F.	.007	.14	.19	.33	
Soffit & fascia, aluminum, vented	.220	L.F.	.032	.46	.86	1.32	
Flashing, aluminum, mill finish, .013" thick	1.500	S.F.	.083	.54	2.58	3.12	
TOTAL			.144	1.59	4.35	5.94	
ASPHALT, ROOF SHINGLES, PREMIUM LAMINATED							
Premium laminated, multi-layered, 260-300 lb./sq.	1.400	S.F.	.032	.90	.90	1.80	
Drip edge, metal, 5" wide	.220	L.F.	.004	.06	.13	.19	
Building paper, #15 asphalt felt	1.500	S.F.	.002	.04	.06	.10	
Ridge shingles, asphalt	.280	L.F.	.007	.14	.19	.33	
Soffit & fascia, aluminum, vented	.220	L.F.	.032	.46	.86	1.32	
Flashing, aluminum, mill finish, .013" thick	1.500	S.F.	.083	.54	2.58	3.12	
TOTAL			.160	2.14	4.72	6.86	
WOOD, CEDAR, NO. 1 PERFECTIONS							
Shingles, red cedar, No.1 perfections, 18" long, 5" exp.	1.400	S.F.	.041	2.77	1.20	3.97	
Drip edge, metal, 5" wide	.220	L.F.	.004	.06	.13	.19	
Building paper, #15 asphalt felt	1.500	S.F.	.002	.04	.06	.10	
Ridge shingles, wood	.280	L.F.	.008	.51	.24	.75	
Soffit & fascia, aluminum, vented	.220	L.F.	.032	.46	.86	1.32	
Flashing, aluminum, mill finish, .013" thick	1.500	S.F.	.083	.54	2.58	3.12	
TOTAL			.170	4.38	5.07	9.45	
SLATE, BUCKINGHAM, BLACK							
Shingles, Buckingham, Virginia, black	1.400	S.F.	.064	7.84	1.81	9.65	
Drip edge, metal, 5" wide	.220	L.F.	.004	.06	.13	.19	
Building paper, #15 asphalt felt	1.500	S.F.	.002	.04	.06	.10	
Ridge shingles, slate	.280	L.F.	.011	1.32	.32	1.64	
Soffit & fascia, aluminum, vented	.220	L.F.	.032	.46	.86	1.32	
Flashing, copper, 16 oz.	1.500	S.F.	.104	5.37	3.26	8.63	
TOTAL			.217	15.09	6.44	21.53	

Gable Dormer Roofing Systems

System Description	QUAN.	UNIT	LABOR-HOURS	COST PER S.F.			S.F. EXTENSIONS
				MAT.	INST.	TOTAL	
WOOD, CEDAR, RESQUARED AND REBUTTED							
Resquared & rebutted, 18" long, 5" exposure	1.400	S.F.	.037	1.54	1.10	2.64	
Drip edge, metal, 5" wide	.220	L.F.	.004	.06	.13	.19	
Building paper, #15 asphalt felt	1.500	S.F.	.002	.04	.06	.10	
Ridge shingles, wood	.280	L.F.	.011	1.32	.32	1.64	
Soffit & fascia, aluminum, vented	.220	L.F.	.032	.46	.86	1.32	
Flashing, aluminum, mill finish, .013" thick	1.500	S.F.	.104	5.37	3.26	8.63	
TOTAL			.190	8.79	5.73	14.52	
WOOD SHAKES							
Shingles, wood shakes hand split, 24" long, 10" exp.	1.400	S.F.	.045	2.31	1.32	3.63	
Drip edge, metal, 5" wide	.220	L.F.	.004	.06	.13	.19	
Building paper, #15 asphalt felt	1.500	S.F.	.002	.04	.06	.10	
Ridge shingles, shakes	.280	L.F.	.011	1.32	.32	1.64	
Soffit & fascia, aluminum, vented	.220	L.F.	.032	.46	.86	1.32	
Flashing, aluminum, mill finish, 0.013" thick	1.500	S.F.	.104	5.37	3.26	8.63	
TOTAL			.198	9.56	5.95	15.51	

The prices in these systems are based on a square foot of plan area under the dormer roof.

Gable Dormer Roofing System Components

Component Description	QUAN.	UNIT	LABOR-HOURS	COST PER S.F.		
				MAT.	INST.	TOTAL
Shingles						
Shingles, asphalt, standard, inorganic, class A, 210-235 lb./sq.	1.400	S.F.	.016	.35	.53	.88
Laminated, multi-layered, 240-260 lb./sq.	1.400	S.F.	.025	.64	.70	1.34
Clay tile, Spanish tile, red	1.400	S.F.	.062	5.95	1.75	7.70
Mission tile, red	1.400	S.F.	.097	9.40	2.74	12.14
French tile, red	1.400	S.F.	.083	8.45	2.34	10.79
Slate Buckingham, Virginia, black	1.400	S.F.	.064	7.85	1.81	9.66
Vermont, black or grey	1.400	S.F.	.064	5.90	1.81	7.71
Wood, red cedar, No.1 5X, 16" long, 5" exposure	1.400	S.F.	.045	2.39	1.32	3.71
Fire retardant	1.400	S.F.	.045	2.94	1.32	4.26
Shakes hand split, 24" long, 10" exposure	1.400	S.F.	.045	2.31	1.32	3.63
18" long, 8" exposure	1.400	S.F.	.056	2.11	1.64	3.75
Flashing						
Flashing, aluminum, .013" thick	1.500	S.F.	.083	.54	2.58	3.12
.032" thick	1.500	S.F.	.083	1.52	2.58	4.10
.040" thick	1.500	S.F.	.083	2.70	2.58	5.28
.050" thick	1.500	S.F.	.083	3.15	2.58	5.73
Copper, 16 oz.	1.500	S.F.	.104	5.35	3.26	8.61
20 oz.	1.500	S.F.	.109	6.70	3.41	10.11
24 oz.	1.500	S.F.	.114	8.05	3.56	11.61
32 oz.	1.500	S.F.	.120	10.75	3.74	14.49

Shed Dormer Roofing Systems

System Description	QUAN.	UNIT	LABOR-HOURS	COST PER S.F.			S.F. EXTENSIONS
				MAT.	INST.	TOTAL	
ASPHALT, ROOF SHINGLES, CLASS A							
Shingles, standard inorganic class A 210-235 lb./sq.	1.100	S.F.	.013	.28	.42	.70	
Drip edge, aluminum, 5" wide	.250	L.F.	.005	.05	.15	.20	
Building paper, #15 asphalt felt	1.200	S.F.	.002	.03	.05	.08	
Soffit & fascia, aluminum, vented, 1' overhang	.250	L.F.	.036	.52	.98	1.50	
Flashing, aluminum, mill finish, 0.013" thick	.800	L.F.	.044	.29	1.38	1.67	
TOTAL			.100	1.17	2.98	4.15	
ASPHALT, ROOF SHINGLES, PREMIUM LAMINATED							
Premium laminated, multi-layered, 260-300 lb./sq.	1.100	S.F.	.025	.71	.71	1.42	
Drip edge, aluminum, 5" wide	.250	L.F.	.005	.05	.15	.20	
Building paper, #15 asphalt felt	1.200	S.F.	.002	.03	.05	.08	
Soffit & fascia, aluminum, vented, 1' overhang	.250	L.F.	.036	.52	.98	1.50	
Flashing, aluminum, mill finish, 0.013" thick	.800	L.F.	.044	.29	1.38	1.67	
TOTAL			.112	1.60	3.27	4.87	
CLAY TILE, SPANISH TILE							
Clay tile, Spanish tile, red	1.100	S.F.	.049	4.68	1.38	6.06	
Drip edge, aluminum, 5" wide	.250	L.F.	.005	.05	.15	.20	
Building paper, #15 asphalt felt	1.200	S.F.	.002	.03	.05	.08	
Soffit & fascia, aluminum, vented, 1' overhang	.250	L.F.	.036	.52	.98	1.50	
Flashing, copper, 16 oz.	.800	L.F.	.056	2.86	1.74	4.60	
TOTAL			.148	8.14	4.30	12.44	
WOOD, CEDAR, NO. 1 PERFECTIONS, 18" LONG							
Shingles, wood, red cedar, #1 perfections, 5" exposure	1.100	S.F.	.032	2.18	.94	3.12	
Drip edge, aluminum, 5" wide	.250	L.F.	.005	.05	.15	.20	
Building paper, #15 asphalt felt	1.200	S.F.	.002	.03	.05	.08	
Soffit & fascia, aluminum, vented, 1' overhang	.250	L.F.	.036	.52	.98	1.50	
Flashing, aluminum, mill finish, 0.013" thick	.800	L.F.	.044	.29	1.38	1.67	
TOTAL			.119	3.07	3.50	6.57	
WOOD, CEDAR, RESQUARED AND REBUTTED, 18" LONG							
Resquared & rebutted, 18" long, 5" exposure	1.100	S.F.	.029	1.21	.86	2.07	
Drip edge, aluminum, 5" wide	.250	L.F.	.005	.05	.15	.20	
Building paper, #15 asphalt felt	1.200	S.F.	.002	.03	.05	.08	
Soffit & fascia, aluminum, vented, 1' overhang	.250	L.F.	.036	.52	.98	1.50	
Flashing, aluminum, mill finish, 0.013" thick	.800	L.F.	.044	.29	1.38	1.67	
TOTAL			.116	2.10	3.42	5.52	

Shed Dormer Roofing Systems

System Description	QUAN.	UNIT	LABOR-HOURS	COST PER S.F.			S.F. EXTENSIONS
				MAT.	INST.	TOTAL	
WOOD SHAKES							
Shingles, wood shakes hand split, 24'' long, 10'' exp.	1.100	S.F.	.035	1.82	1.03	2.85	
Drip edge, metal 5'' wide	.250	L.F.	.005	.05	.15	.20	
Building paper, #15 asphalt felt	1.200	S.F.	.002	.03	.05	.08	
Soffit & fascia, aluminum, vented, 1' overhang	.250	L.F.	.036	.52	.98	1.50	
Flashing, aluminum, mill finish, 0.013'' thick	.800	L.F.	.044	.29	1.38	1.67	
TOTAL			.122	2.71	3.59	6.30	

The prices in this system are based on a square foot of plan area under the dormer roof.

Shed Dormer Roofing System Components

Component Description	QUAN.	UNIT	LABOR-HOURS	COST PER S.F.		
				MAT.	INST.	TOTAL
Shingles						
Shingles, asphalt, standard, inorganic, class A, 210-235 lb./sq.	1.100	S.F.	.013	.28	.42	.70
Laminated, multi-layered, 240-260 lb./sq.	1.100	S.F.	.020	.50	.55	1.05
Premium laminated, multi-layered, 260-300 lb./sq.	1.100	S.F.	.025	.71	.71	1.42
Clay tile, Spanish tile, red	1.100	S.F.	.049	4.68	1.38	6.06
Mission tile, red	1.100	S.F.	.077	7.35	2.16	9.51
French tile, red	1.100	S.F.	.065	6.65	1.84	8.49
Slate Buckingham, Virginia, black	1.100	S.F.	.050	6.15	1.42	7.57
Vermont, black or grey	1.100	S.F.	.050	4.62	1.42	6.04
Wood, red cedar, No. 1 5X, 16'' long, 5'' exposure	1.100	S.F.	.035	1.88	1.03	2.91
Fire retardant	1.100	S.F.	.035	2.31	1.03	3.34
Shakes hand split, 24'' long, 10'' exposure	1.100	S.F.	.035	1.82	1.03	2.85
Fire retardant	1.100	S.F.	.035	2.28	1.03	3.31
Flashing						
Flashing, aluminum, .013'' thick	.800	L.F.	.044	.29	1.38	1.67
.032'' thick	.800	L.F.	.044	.81	1.38	2.19
.040'' thick	.800	L.F.	.044	1.44	1.38	2.82
.050'' thick	.800	L.F.	.044	1.68	1.38	3.06
Copper, 16 oz.	.800	L.F.	.056	2.86	1.74	4.60
20 oz.	.800	L.F.	.058	3.58	1.82	5.40
24 oz.	.800	L.F.	.061	4.28	1.90	6.18
32 oz.	.800	L.F.	.064	5.70	1.99	7.69

Skylight/Skywindow Systems

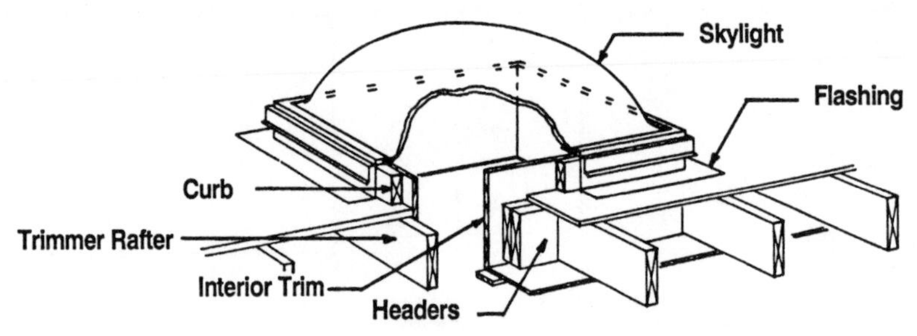

System Description	QUAN.	UNIT	LABOR-HOURS	COST EACH			EA. EXTENSIONS
				MAT.	INST.	TOTAL	
SKYLIGHT, FIXED, 32" X 32"							
Skylight, fixed bubble, insulating, 32" x 32"	1.000	Ea.	1.422	128.71	38.68	167.39	
Trimmer rafters, 2" x 6"	28.000	L.F.	.448	16.24	12.04	28.28	
Headers, 2" x 6"	6.000	L.F.	.267	3.48	7.20	10.68	
Curb, 2" x 4"	12.000	L.F.	.154	4.80	4.20	9.00	
Flashing, aluminum, .013" thick	13.500	S.F.	.745	4.86	23.22	28.08	
Trim, interior casing, painted	12.000	L.F.	.617	9.00	17.40	26.40	
TOTAL			3.653	167.09	102.74	269.83	
SKYLIGHT, FIXED, 48" X 48"							
Skylight, fixed bubble, insulating, 48" x 48"	1.000	Ea.	1.296	230.40	35.36	265.76	
Trimmer rafters, 2" x 6"	28.000	L.F.	.448	16.24	12.04	28.28	
Headers, 2" x 6"	8.000	L.F.	.356	4.64	9.60	14.24	
Curb, 2" x 4"	16.000	L.F.	.205	6.40	5.60	12.00	
Flashing, aluminum, .013" thick	16.000	S.F.	.883	5.76	27.52	33.28	
Trim, interior casing, painted	16.000	L.F.	.821	12.00	23.20	35.20	
TOTAL			4.009	275.44	113.32	388.76	
SKYLIGHT, VENTILATING, 36" X 36"							
Skylight, ventilating bubble insulating, 36" x 36"	1.000	Ea.	2.667	350.00	72.53	422.53	
Trimmer rafters, 2" x 6"	28.000	L.F.	.448	16.24	12.04	28.28	
Headers, 2" x 6"	6.000	L.F.	.356	4.64	9.60	14.24	
Curb, 2" x 4"	12.000	L.F.	.205	6.40	5.60	12.00	
Flashing, aluminum, .013" thick	13.500	S.F.	.883	5.76	27.52	33.28	
Trim, interior casing, painted	12.000	L.F.	.821	12.00	23.20	35.20	
TOTAL			5.380	395.04	150.49	545.53	
SKYLIGHT, VENTILATING, 36" X 52"							
Skylight, ventilating bubble insulating, 36" x 52"	1.000	Ea.	3.200	445.00	87.04	532.04	
Trimmer rafters, 2" x 6"	28.000	L.F.	.448	16.24	12.04	28.28	
Headers, 2" x 6"	8.000	L.F.	.356	4.64	9.60	14.24	
Curb, 2" x 4"	16.000	L.F.	.205	6.40	5.60	12.00	
Flashing, aluminum, .013" thick	16.000	S.F.	.883	5.76	27.52	33.28	
Trim, interior casing, painted	16.000	L.F.	.821	12.00	23.20	35.20	
TOTAL			5.913	490.04	165.00	655.04	

Skylight/Skywindow Systems

System Description	QUAN.	UNIT	LABOR-HOURS	COST EACH			EA. EXTENSIONS
				MAT.	INST.	TOTAL	
SKYWINDOW, OPERATING, 24″ X 48″							
Skywindow, operating, thermopane glass, 24″ x 48″	1.000	Ea.	3.200	515.00	87.04	602.04	
Trimmer rafters, 2″ x 6″	28.000	L.F.	.448	16.24	12.04	28.28	
Headers, 2″ x 6″	8.000	L.F.	.267	3.48	7.20	10.68	
Curb, 2″ x 4″	14.000	L.F.	.179	5.60	4.90	10.50	
Flashing, aluminum, .013″ thick	14.000	S.F.	.772	5.04	24.08	29.12	
Trim, interior casing, painted	14.000	L.F.	.719	10.50	20.30	30.80	
TOTAL			5.585	555.86	155.56	711.42	
SKYWINDOW, OPERATING, 32″ X 48″							
Skywindow, operating, thermopane glass, 32″ x 48″	1.000	Ea.	3.556	535.00	96.88	631.88	
Trimmer rafters, 2″ x 6″	28.000	L.F.	.448	16.24	12.04	28.28	
Headers, 2″ x 6″	8.000	L.F.	.356	4.64	9.60	14.24	
Curb, 2″ x 4″	14.000	L.F.	.179	5.60	4.90	10.50	
Flashing, aluminum, .013″ thick	14.000	S.F.	.772	5.04	24.08	29.12	
Trim, interior casing, painted	14.000	L.F.	.719	10.50	20.30	30.80	
TOTAL			6.030	577.02	167.80	744.82	

The prices in these systems are on a cost each basis.

Skylight/Skywindow System Components

Component Description	QUAN.	UNIT	LABOR-HOURS	COST EACH		
				MAT.	INST.	TOTAL
Skylights						
Skylight, fixed bubble insulating, 24″ x 24″	1.000	Ea.	.800	72.50	22.00	94.50
32″ x 48″	1.000	Ea.	.864	154.00	23.50	177.50
Ventilating bubble insulating, 36″ x 36″	1.000	Ea.	2.667	350.00	72.50	422.50
52″ x 52″	1.000	Ea.	2.667	525.00	72.50	597.50
28″ x 52″	1.000	Ea.	3.200	410.00	87.00	497.00
36″ x 52″	1.000	Ea.	3.200	445.00	87.00	532.00
Trimer rafters						
Trimmer rafters, 2″ x 6″	28.000	L.F.	.448	16.25	12.05	28.30
2″ x 8″	28.000	L.F.	.472	23.00	12.60	35.60
2″ x 10″	28.000	L.F.	.711	33.50	19.05	52.55
Headers						
Headers, 24″ window, 2″ x 6″	4.000	L.F.	.178	2.32	4.80	7.12
2″ x 8″	4.000	L.F.	.188	3.32	5.10	8.42
2″ x 10″	4.000	L.F.	.200	4.80	5.35	10.15
32″ window, 2″ x 6″	6.000	L.F.	.267	3.48	7.20	10.68
2″ x 8″	6.000	L.F.	.282	4.98	7.60	12.58
2″ x 10″	6.000	L.F.	.300	7.20	8.05	15.25
48″ window, 2″ x 6″	8.000	L.F.	.356	4.64	9.60	14.24
2″ x 8″	8.000	L.F.	.376	6.65	10.20	16.85
2″ x 10″	8.000	L.F.	.400	9.60	10.75	20.35

Interior Systems

Drywall & Thincoat Wall Systems

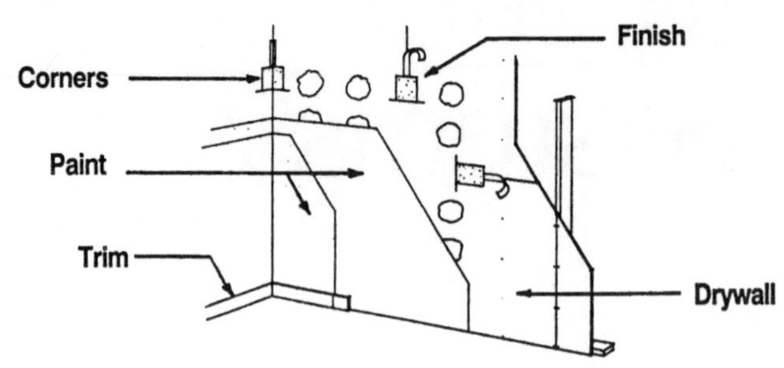

System Description	QUAN.	UNIT	LABOR-HOURS	COST PER S.F.			S.F. EXTENSIONS
				MAT.	INST.	TOTAL	
1/2" SHEETROCK, TAPED & FINISHED							
Drywall, 1/2" thick, standard	1.000	S.F.	.008	.17	.23	.40	
Finish, taped & finished joints	1.000	S.F.	.008	.09	.23	.32	
Corners, taped & finished, 32 L.F. per 12' x 12' room	.083	L.F.	.001	.01	.04	.05	
Painting, primer & 2 coats	1.000	S.F.	.010	.12	.26	.38	
Trim, baseboard, painted	.125	L.F.	.006	.15	.17	.32	
TOTAL			.033	.54	.93	1.47	
FIRE RESISTANT, 1/2" SHEETROCK, TAPED & FINISHED							
Drywall, 1/2" thick, fire resistant	1.000	S.F.	.008	.23	.23	.46	
Finish, taped & finished joints	1.000	S.F.	.008	.09	.23	.32	
Corners, taped & finished, 32 L.F. per 12' x 12' room	.083	L.F.	.001	.01	.04	.05	
Painting, primer & 2 coats	1.000	S.F.	.010	.12	.26	.38	
Trim, baseboard, painted	.125	L.F.	.006	.15	.17	.32	
TOTAL			.033	.60	.93	1.53	
WATER RESISTANT, 1/2" SHEETROCK, TAPED & FINISHED							
Drywall, 1/2" thick, water resistant	1.000	S.F.	.008	.25	.23	.48	
Finish, taped & finished joints	1.000	S.F.	.008	.09	.23	.32	
Corners, taped & finished, 32 L.F. per 12' x 12' room	.083	L.F.	.001	.01	.04	.05	
Painting, primer & 2 coats	1.000	S.F.	.010	.12	.26	.38	
Trim, baseboard, painted	.125	L.F.	.006	.15	.17	.32	
TOTAL			.033	.62	.93	1.55	
THINCOAT, SKIM-COAT, ON 1/2" BACKER DRYWALL							
Drywall, 1/2" thick, thincoat backer	1.000	S.F.	.008	.17	.23	.40	
Thincoat plaster	1.000	S.F.	.011	.12	.29	.41	
Corners, taped & finished, 32 L.F. per 12' x 12' room	.083	L.F.	.001	.01	.04	.05	
Painting, primer & 2 coats	1.000	S.F.	.010	.12	.26	.38	
Trim, baseboard, painted	.125	L.F.	.006	.15	.17	.32	
TOTAL			.036	.57	.99	1.56	
5/8" SHEETROCK, TAPED & FINISHED							
Drywall, 5/8" thick, standard	1.000	S.F.	.008	.23	.23	.46	
Finish, taped & finished joints	1.000	S.F.	.008	.09	.23	.32	
Corners, taped & finished, 32 L.F. per 12' x 12' room	.083	L.F.	.001	.01	.04	.05	
Painting, primer & 2 coats	1.000	S.F.	.010	.12	.26	.38	
Trim, baseboard, painted	.125	L.F.	.006	.15	.17	.32	
TOTAL			.033	.60	.93	1.53	

Drywall & Thincoat Wall Systems

System Description	QUAN.	UNIT	LABOR-HOURS	COST PER S.F.			S.F. EXTENSIONS
				MAT.	INST.	TOTAL	
3/8" SHEETROCK, TAPED & FINISHED							
Drywall, 3/8" thick, standard	1.000	S.F.	.008	.17	.23	.40	
Finish, taped & finished joints	1.000	S.F.	.008	.09	.23	.32	
Corners, taped & finished, 32 L.F. per 12' x 12' room	.083	L.F.	.001	.01	.04	.05	
Painted, primer & 2 coats	1.000	S.F.	.010	.12	.26	.38	
Trim, baseboard, painted	.125	L.F.	.006	.15	.17	.32	
TOTAL			.033	.54	.93	1.47	
THINCOAT, SKIM-COAT, ON 3/8" BACKER DRYWALL							
Drywall, 3/8" thick, thincoat backer	1.000	S.F.	.008	.17	.23	.40	
Thincoat, plaster	1.000	S.F.	.011	.12	.29	.41	
Corners, taped & finished, 32 L.F. per 12' x 12' room	.083	L.F.	.001	.01	.04	.05	
Painting, primer & 2 coats	1.000	S.F.	.010	.12	.26	.38	
Trim, baseboard, painted	.125	L.F.	.006	.15	.17	.32	
TOTAL			.036	.57	.99	1.56	
WATER-RESISTANT, 3/8" SHEETROCK, TAPED & FINISHED							
Drywall, 3/8" thick, water-resistant	1.000	S.F.	.008	.20	.23	.43	
Finish, taped & finished joints	1.000	S.F.	.008	.09	.23	.32	
Corners, taped & finished, 32 L.F. per 12' x 12' room	.083	L.F.	.001	.01	.04	.05	
Painting, primer & 2 coats	1.000	S.F.	.010	.12	.26	.38	
Trim, baseboard, painted	.125	L.F.	.006	.15	.17	.32	
TOTAL			.033	.57	.93	1.50	

The costs in this system are based on a square foot of wall.
Do not deduct for openings.

Drywall & Thincoat Wall System Components

Component Description	QUAN.	UNIT	LABOR-HOURS	COST PER S.F.		
				MAT.	INST.	TOTAL
Drywall - sheetrock						
Drywall-sheetrock, 1/2" thick, standard	1.000	S.F.	.008	.17	.23	.40
Fire resistant	1.000	S.F.	.008	.23	.23	.46
Water resistant	1.000	S.F.	.008	.25	.23	.48
Paneling, not including furring or trim						
Plywood, prefinished, 1/4" thick, 4' x 8' sheets, vert. grooves						
Birch faced, minimum	1.000	S.F.	.032	.77	.86	1.63
Average	1.000	S.F.	.038	1.10	1.03	2.13
Maximum	1.000	S.F.	.046	1.60	1.23	2.83
Mahogany, African	1.000	S.F.	.040	2.04	1.08	3.12
Philippine (lauan)	1.000	S.F.	.032	.88	.86	1.74
Oak or cherry, minimum	1.000	S.F.	.032	1.70	.86	2.56
Maximum	1.000	S.F.	.040	2.86	1.08	3.94
Rosewood	1.000	S.F.	.050	4.07	1.34	5.41
Teak	1.000	S.F.	.040	2.86	1.08	3.94
Chestnut	1.000	S.F.	.043	4.23	1.14	5.37
Pecan	1.000	S.F.	.040	1.82	1.08	2.90
Walnut, minimum	1.000	S.F.	.032	2.42	.86	3.28
Maximum	1.000	S.F.	.040	4.62	1.08	5.70

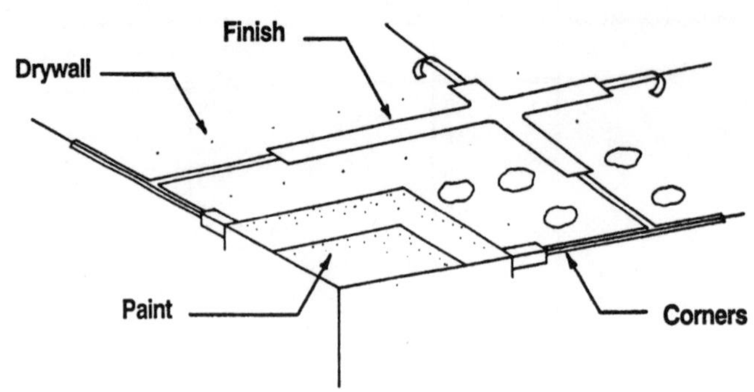

System Description	QUAN.	UNIT	LABOR-HOURS	COST PER S.F.			S.F. EXTENSIONS
				MAT.	INST.	TOTAL	
1/2'' SHEETROCK, TAPED & FINISHED							
Drywall, 1/2'' thick, standard	1.000	S.F.	.008	.17	.23	.40	
Finish, taped & finished	1.000	S.F.	.008	.09	.23	.32	
Corners, taped & finished, 12' x 12' room	.333	L.F.	.005	.02	.14	.16	
Paint, primer & 2 coats	1.000	S.F.	.010	.12	.26	.38	
TOTAL			.031	.40	.86	1.26	
THINCOAT, SKIM COAT ON 1/2'' BACKER DRYWALL							
Drywall, 1/2'' thick, thincoat backer	1.000	S.F.	.008	.17	.23	.40	
Thincoat plaster	1.000	S.F.	.011	.12	.29	.41	
Corners, taped & finished, 12' x 12' room	.333	L.F.	.005	.02	.14	.16	
Paint, primer & 2 coats	1.000	S.F.	.010	.12	.26	.38	
TOTAL			.034	.43	.92	1.35	
FIRE RESISTANT, 1/2'' SHEETROCK, TAPED & FINISHED							
Drywall, 1/2'' thick, fire resistant	1.000	S.F.	.008	.23	.23	.46	
Finish, taped & finished	1.000	S.F.	.008	.09	.23	.32	
Corners, taped & finished, 12' x 12' room	.333	L.F.	.005	.02	.14	.16	
Paint, primer & 2 coats	1.000	S.F.	.010	.12	.26	.38	
TOTAL			.031	.46	.86	1.32	
WATER-RESISTANT SHEETROCK, 1/2'' THICK, TAPED & FINISHED							
Drywall, 1/2'' thick, water-resistant	1.000	S.F.	.008	.25	.23	.48	
Finish, taped & finished	1.000	S.F.	.008	.09	.23	.32	
Corners, taped & finished, 12' x 12' room	.333	L.F.	.005	.02	.14	.16	
Paint, primer & 2 coats	1.000	S.F.	.010	.12	.26	.38	
TOTAL			.031	.48	.86	1.34	
5/8'' SHEETROCK, TAPED & FINISHED							
Drywall, 5/8'' thick, standard	1.000	S.F.	.008	.23	.23	.46	
Finish, taped & finished	1.000	S.F.	.008	.09	.23	.32	
Corners, taped & finished, 12' x 12' room	.333	L.F.	.005	.02	.14	.16	
Paint, primer & 2 coats	1.000	S.F.	.010	.12	.26	.38	
TOTAL			.031	.46	.86	1.32	

Drywall & Thincoat Ceiling Systems

System Description	QUAN.	UNIT	LABOR-HOURS	COST PER S.F.			S.F. EXTENSIONS
				MAT.	INST.	TOTAL	
3/8" SHEETROCK, TAPED & FINISHED							
Drywall, 3/8" thick, standard	1.000	S.F.	.008	.17	.23	.40	
Finish, taped & finished	1.000	S.F.	.008	.09	.23	.32	
Corners, taped & finished, 12' x 12' room	.330	L.F.	.005	.02	.14	.16	
Paint, primer & 2 coats	1.000	S.F.	.010	.12	.26	.38	
TOTAL			.031	.40	.86	1.26	
THINCOAT, SKIM-COAT ON 3/8" BACKER DRYWALL							
Drywall, 3/8" thick, thincoat backer	1.000	S.F.	.008	.17	.23	.40	
Thincoat plaster	1.000	S.F.	.011	.12	.29	.41	
Corners, taped & finished, 12' x 12' room	.330	L.F.	.005	.02	.14	.16	
Paint, primer & 2 coats	1.000	S.F.	.010	.12	.26	.38	
TOTAL			.034	.43	.92	1.35	
WATER-RESISTANT, 3/8" SHEETROCK, TAPED & FINISHED							
Drywall, 3/8" thick, water-resistant	1.000	S.F.	.008	.20	.23	.43	
Finish, taped & finished	1.000	S.F.	.008	.09	.23	.32	
Corners, taped & finished, 12' x 12' room	.330	L.F.	.005	.02	.14	.16	
Paint, primer & 2 coats	1.000	S.F.	.010	.12	.26	.38	
TOTAL			.031	.43	.86	1.29	

The costs in this system are based on a square foot of ceiling.

Drywall & Thincoat Ceiling System Components

Component Description	QUAN.	UNIT	LABOR-HOURS	COST PER S.F.		
				MAT.	INST.	TOTAL
Drywall - sheetrock						
5/8" thick, standard	1.000	S.F.	.008	.23	.23	.46
Fire resistant	1.000	S.F.	.008	.23	.23	.46
Water resistant	1.000	S.F.	.008	.28	.23	.51
Drywall backer for thincoat system, 1/2" thick	1.000	S.F.	.016	.40	.46	.86
5/8" thick	1.000	S.F.	.016	.46	.46	.92
Corners taped & finished						
Corners taped & finished, 4' x 4' room	1.000	L.F.	.015	.07	.43	.50
6' x 6' room	.667	L.F.	.010	.05	.29	.34
10' x 10' room	.400	L.F.	.006	.03	.17	.20
12' x 12' room	.333	L.F.	.005	.02	.14	.16
16' x 16' room	.250	L.F.	.003	.01	.08	.09
Thincoat system, 4' x 4' room	1.000	L.F.	.011	.12	.29	.41
6' x 6' room	.667	L.F.	.007	.08	.20	.28
10' x 10' room	.400	L.F.	.004	.05	.11	.16
12' x 12' room	.333	L.F.	.004	.04	.09	.13
16' x 16' room	.250	L.F.	.002	.02	.05	.07
Tile						
Tile, ceramic adhesive thin set, 4 1/4" x 4 1/4" tiles	1.000	S.F.	.084	2.11	2.08	4.19
6" x 6" tiles	1.000	S.F.	.080	2.46	1.98	4.44
Pregrouted sheets	1.000	S.F.	.067	3.85	1.65	5.50

Plaster & Stucco Wall Systems

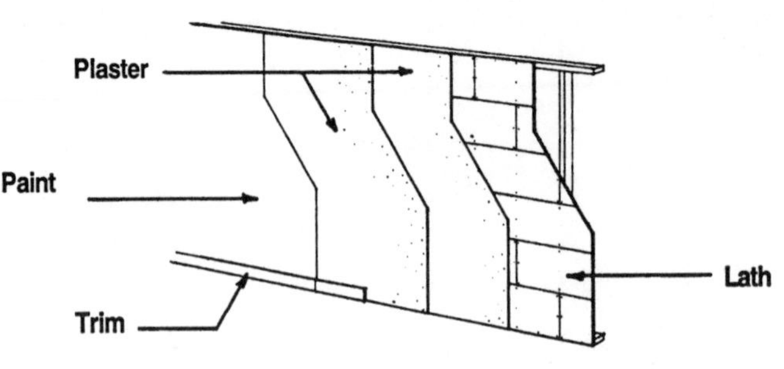

System Description	QUAN.	UNIT	LABOR-HOURS	COST PER S.F.			S.F. EXTENSIONS
				MAT.	INST.	TOTAL	
PLASTER ON GYPSUM LATH							
Plaster, gypsum or perlite, 2 coats	1.000	S.F.	.042	.40	1.13	1.53	
Lath, 3/8'' gypsum	1.000	S.F.	.010	.43	.29	.72	
Corners, expanded metal, 32 L.F. per 12' x 12' room	.083	L.F.	.002	.01	.05	.06	
Painting, primer & 2 coats	1.000	S.F.	.010	.12	.26	.38	
Trim, baseboard, painted	.125	L.F.	.006	.15	.17	.32	
TOTAL			.070	1.11	1.90	3.01	
PLASTER ON GYPSUM LATH, 3 COATS							
Plaster, gypsum or perlite, 3 coats	1.000	S.F.	.051	.55	1.37	1.92	
Lath, 3/8'' gypsum	1.000	S.F.	.010	.43	.29	.72	
Corners, expanded metal, 32 L.F. per 12' x 12' room	.083	L.F.	.002	.01	.05	.06	
Painting, primer & 2 coats	1.000	S.F.	.010	.12	.26	.38	
Trim, baseboard, painted	.125	L.F.	.006	.15	.17	.32	
TOTAL			.079	1.26	2.14	3.40	
PLASTER ON METAL LATH							
Plaster, gypsum or perlite, 2 coats	1.000	S.F.	.042	.40	1.13	1.53	
Lath, 2.5 Lb. diamond, metal	1.000	S.F.	.010	.19	.29	.48	
Corners, expanded metal, 32 L.F. per 12' x 12' room	.083	L.F.	.002	.01	.05	.06	
Painting, primer & 2 coats	1.000	S.F.	.010	.12	.26	.38	
Trim, baseboard, painted	.125	L.F.	.006	.15	.17	.32	
TOTAL			.070	.87	1.90	2.77	
PLASTER ON METAL LATH, 3 COATS							
Plaster, gypsum or perlite, 3 coats	1.000	S.F.	.051	.55	1.37	1.92	
Lath, 2.5 Lb. diamond, metal	1.000	S.F.	.010	.19	.29	.48	
Corners, expanded metal, 32 L.F. per 12' x 12' room	.083	L.F.	.002	.01	.05	.06	
Painting, primer & 2 coats	1.000	S.F.	.010	.12	.26	.38	
Trim, baseboard, painted	.125	L.F.	.006	.15	.17	.32	
TOTAL			.079	1.02	2.14	3.16	
STUCCO ON METAL LATH							
Stucco, 2 coats	1.000	S.F.	.041	.32	1.08	1.40	
Lath, 2.5 Lb. diamond, metal	1.000	S.F.	.010	.19	.29	.48	
Corners, expanded metal, 32 L.F. per 12' x 12' room	.083	L.F.	.002	.01	.05	.06	
Painting, primer & 2 coats	1.000	S.F.	.010	.12	.26	.38	
Trim, baseboard, painted	.125	L.F.	.006	.15	.17	.32	
TOTAL			.069	.79	1.85	2.64	

Plaster & Stucco Wall Systems

System Description	QUAN.	UNIT	LABOR-HOURS	COST PER S.F.			S.F. EXTENSIONS
				MAT.	INST.	TOTAL	
STUCCO ON METAL LATH, 3 COATS							
Stucco, 3 coats	1.000	S.F.	.102	.69	2.74	3.43	
Lath, 2.5 Lb. diamond, metal	1.000	S.F.	.010	.19	.29	.48	
Corners, expanded metal, 32 L.F. per 12' x 12' room	.083	L.F.	.002	.01	.05	.06	
Painting, primer & 2 coats	1.000	S.F.	.010	.12	.26	.38	
Trim, baseboard, painted	.125	L.F.	.006	.15	.17	.32	
TOTAL			.130	1.16	3.51	4.67	

The costs in these systems are based on a per square foot of wall area.
Do not deduct for openings.

Plaster & Stucco Wall System Components

Component Description	QUAN.	UNIT	LABOR-HOURS	COST PER S.F.		
				MAT.	INST.	TOTAL
Lath						
Lath, gypsum, standard, 3/8'' thick	1.000	S.F.	.010	.43	.29	.72
1/2'' thick	1.000	S.F.	.013	.50	.35	.85
Fire resistant, 3/8'' thick	1.000	S.F.	.013	.48	.35	.83
1/2'' thick	1.000	S.F.	.014	.52	.38	.90
Metal, diamond, 2.5 Lb.	1.000	S.F.	.010	.19	.29	.48
3.4 Lb.	1.000	S.F.	.012	.24	.33	.57
Rib, 2.75 Lb.	1.000	S.F.	.012	.18	.33	.51
3.4 Lb.	1.000	S.F.	.013	.33	.35	.68
Corners						
Corners, expanded metal, 32 L.F. per 4' x 4' room	.250	L.F.	.005	.03	.15	.18
6' x 6' room	.110	L.F.	.002	.01	.06	.07
10' x 10' room	.100	L.F.	.002	.01	.06	.07
16' x 16' room	.063	L.F.	.001	.01	.04	.05
Painting						
Painting, primer & 1 coats	1.000	S.F.	.007	.08	.18	.26
Wallpaper, $7.00/double roll	1.000	S.F.	.013	.28	.33	.61
$17.00/double roll	1.000	S.F.	.015	.47	.40	.87
$40.00/double roll	1.000	S.F.	.018	.91	.49	1.40
Tile, ceramic thin set, 4-1/4'' x 4-1/4'' tiles	1.000	S.F.	.084	2.11	2.08	4.19
6'' x 6'' tiles	1.000	S.F.	.080	2.46	1.98	4.44
Pregrouted sheets	1.000	S.F.	.067	3.85	1.65	5.50

Plaster & Stucco Ceiling Systems

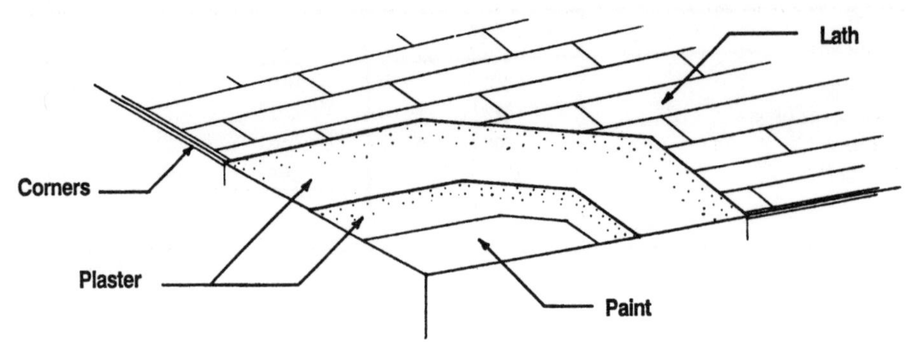

System Description	QUAN.	UNIT	LABOR-HOURS	COST PER S.F.			S.F. EXTENSIONS
				MAT.	INST.	TOTAL	
PLASTER ON GYPSUM LATH							
Plaster, gypsum or perlite, 2 coats	1.000	S.F.	.048	.40	1.29	1.69	
Lath, 3/8'' gypsum	1.000	S.F.	.014	.43	.40	.83	
Corners, expanded metal, 12' x 12' room	.330	L.F.	.007	.03	.19	.22	
Painting, primer & 2 coats	1.000	S.F.	.010	.12	.26	.38	
TOTAL			.079	.98	2.14	3.12	
PLASTER ON GYPSUM LATH, 3 COATS							
Plaster, gypsum or perlite, 3 coats	1.000	S.F.	.051	.55	1.37	1.92	
Lath, 3/8'' gypsum	1.000	S.F.	.014	.43	.40	.83	
Corners, expanded metal, 12' x 12' room	.330	L.F.	.007	.03	.19	.22	
Painting, primer & 2 coats	1.000	S.F.	.010	.12	.26	.38	
TOTAL			.082	1.13	2.22	3.35	
PLASTER ON METAL LATH							
Plaster, gypsum or perlite, 2 coats	1.000	S.F.	.048	.40	1.29	1.69	
Lath, 2.5 Lb. diamond, metal	1.000	S.F.	.012	.19	.33	.52	
Corners, expanded metal, 12' x 12' room	.330	L.F.	.007	.03	.19	.22	
Painting, primer & 2 coats	1.000	S.F.	.010	.12	.26	.38	
TOTAL			.077	.74	2.07	2.81	
PLASTER ON METAL LATH, 3 COATS							
Plaster, gypsum or perlite, 3 coats	1.000	S.F.	.051	.55	1.37	1.92	
Lath, 2.5 Lb. diamond, metal	1.000	S.F.	.012	.19	.33	.52	
Corners, expanded metal, 12' x 12' room	.330	L.F.	.007	.03	.19	.22	
Painting, primer & 2 coats	1.000	S.F.	.010	.12	.26	.38	
TOTAL			.080	.89	2.15	3.04	
STUCCO ON GYPSUM LATH							
Stucco, 2 coats	1.000	S.F.	.041	.32	1.08	1.40	
Lath, 3/8'' gypsum	1.000	S.F.	.014	.43	.40	.83	
Corners, expanded metal, 12' x 12' room	.330	L.F.	.007	.03	.19	.22	
Painting, primer & 2 coats	1.000	S.F.	.010	.12	.26	.38	
TOTAL			.072	.90	1.93	2.83	

Plaster & Stucco Ceiling Systems

System Description	QUAN.	UNIT	LABOR-HOURS	COST PER S.F.			S.F. EXTENSIONS
				MAT.	INST.	TOTAL	
STUCCO ON GYPSUM LATH, 3 COATS							
Stucco, 3 coats	1.000	S.F.	.102	.69	2.74	3.43	
Lath, 3/8'' gypsum	1.000	S.F.	.014	.43	.40	.83	
Corners, expanded metal, 12' x 12' room	.330	L.F.	.007	.03	.19	.22	
Painting, primer & 2 coats	1.000	S.F.	.010	.12	.26	.38	
TOTAL			.133	1.27	3.59	4.86	
STUCCO ON METAL LATH							
Stucco, 2 coats	1.000	S.F.	.041	.32	1.08	1.40	
Lath, 2.5 Lb. diamond, metal	1.000	S.F.	.012	.19	.33	.52	
Corners, expanded metal, 12' x 12' room	.330	L.F.	.007	.03	.19	.22	
Painting, primer & 2 coats	1.000	S.F.	.010	.12	.26	.38	
TOTAL			.070	.66	1.86	2.52	
STUCCO ON METAL LATH, 3 COATS							
Stucco, 3 coats	1.000	S.F.	.102	.69	2.74	3.43	
Lath, 2.5 Lb. diamond, metal	1.000	S.F.	.012	.19	.33	.52	
Corners, expanded metal, 12' x 12' room	.330	L.F.	.007	.03	.19	.22	
Painting, primer & 2 coats	1.000	S.F.	.010	.12	.26	.38	
TOTAL			.131	1.03	3.52	4.55	

The costs in these systems are based on a square foot of ceiling area.

Plaster & Stucco Ceiling System Components

Component Description	QUAN.	UNIT	LABOR-HOURS	COST PER S.F.		
				MAT.	INST.	TOTAL
Lath						
Lath, gypsum, standard, 3/8'' thick	1.000	S.F.	.014	.43	.40	.83
1/2'' thick	1.000	S.F.	.015	.45	.42	.87
Fire resistant, 3/8'' thick	1.000	S.F.	.017	.48	.46	.94
1/2'' thick	1.000	S.F.	.018	.52	.49	1.01
Metal, diamond, 2.5 Lb.	1.000	S.F.	.012	.19	.33	.52
3.4 Lb.	1.000	S.F.	.015	.24	.41	.65
Rib, 2.75 Lb.	1.000	S.F.	.012	.18	.33	.51
3.4 Lb.	1.000	S.F.	.013	.33	.35	.68
Corners						
Corners expanded metal, 4' x 4' room	1.000	L.F.	.020	.10	.59	.69
6' x 6' room	.667	L.F.	.013	.07	.40	.47
10' x 10' room	.400	L.F.	.008	.04	.24	.28
16' x 16' room	.250	L.F.	.004	.02	.11	.13
Painting						
Painting, primer & 1 coat	1.000	S.F.	.007	.08	.18	.26

Interior Door Systems

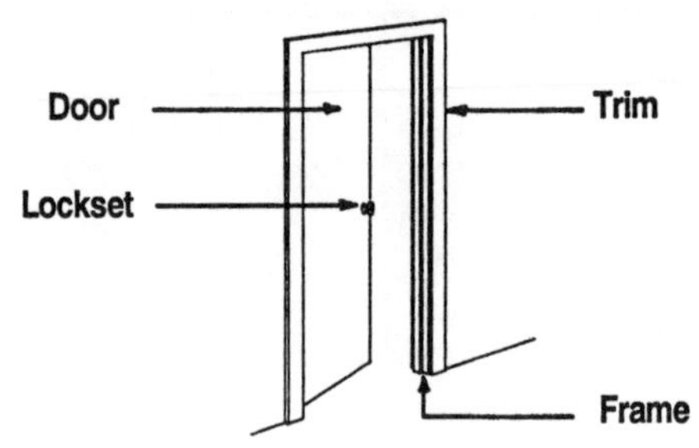

System Description	QUAN.	UNIT	LABOR-HOURS	COST EACH			EA. EXTENSIONS
				MAT.	INST.	TOTAL	
LAUAN, DOOR, HOLLOW CORE, 1-3/8″, 2′-0″ X 6′-8″							
Door, flush lauan, hollow core, 2′-0″ x 6′-8″ high	1.000	Ea.	.889	40.50	24.13	64.63	
Frame, pine, 4-5/8″ jamb	16.000	L.F.	.683	69.76	18.24	88.00	
Trim, casing, painted	32.000	L.F.	1.351	25.92	38.72	64.64	
Butt hinges, chrome, 3-1/2″ x 3-1/2″	1.500	Pr.		23.85		23.85	
Lockset, passage	1.000	Ea.	.500	12.45	14.70	27.15	
Paint, door & frame, primer & 2 coats	2.000	Face	2.353	11.17	62.24	73.41	
TOTAL			5.776	183.65	158.03	341.68	
BIRCH, DOOR, HOLLOW CORE, 1-3/8″, 2′-0″ X 6′-8″							
Door, flush, birch, hollow core, 2′-0″ x 6′-8″ high	1.000	Ea.	.889	48.50	24.13	72.63	
Frame, pine, 4-5/8″ jamb	16.000	L.F.	.683	69.76	18.24	88.00	
Trim, casing, painted	32.000	L.F.	1.351	25.92	38.72	64.64	
Butt hinges, chrome, 3-1/2″ x 3-1/2″	1.500	Pr.		23.85		23.85	
Lockset, passage	1.000	Ea.	.500	12.45	14.70	27.15	
Paint, door & frame, primer & 2 coats	2.000	Face	2.353	11.17	62.24	73.41	
TOTAL			5.776	191.65	158.03	349.68	
PINE, PANELED, 1-3/8″ THICK, 2′-0″ X 6′-8″							
Door, pine, raised panel, 2′-0″ wide x 6′-8″ high	1.000	Ea.	.889	130.00	24.13	154.13	
Frame, pine, 4-5/8″ jamb	16.000	L.F.	.683	69.76	18.24	88.00	
Trim, casing, painted	32.000	L.F.	1.351	25.92	38.72	64.64	
Butt hinges, bronze, 3-1/2″ x 3-1/2″	1.500	Pr.		23.85		23.85	
Lockset, passage	1.000	Ea.	.500	12.45	14.70	27.15	
Paint, door & frame, primer & 2 coats	2.000	Face	2.353	11.17	62.24	73.41	
TOTAL			5.776	273.15	158.03	431.18	
LAUAN, FLUSH DOOR, HOLLOW CORE							
Door, flush, lauan, hollow core, 2′-8″ wide x 6′-8″ high	1.000	Ea.	.889	44.50	24.13	68.63	
Frame, pine, 4-5/8″ jamb	17.000	L.F.	.725	74.12	19.38	93.50	
Trim, casing, painted	34.000	L.F.	1.380	29.58	39.78	69.36	
Butt hinges, chrome, 3-1/2″ x 3-1/2″	1.500	Pr.		23.85		23.85	
Lockset, passage	1.000	Ea.	.500	12.45	14.70	27.15	
Paint, door & frame, primer & 2 coats	2.000	Face	2.465	9.72	64.80	74.52	
TOTAL			5.959	194.22	162.79	357.01	

Interior Door Systems

System Description	QUAN.	UNIT	LABOR-HOURS	MAT.	INST.	TOTAL	EA. EXTENSIONS
BIRCH, FLUSH DOOR, HOLLOW CORE							
Door, flush, birch, hollow core, 2'-8'' wide x 6'-8'' high	1.000	Ea.	.889	55.50	24.13	79.63	
Frame, pine, 4-5/8'' jamb	17.000	L.F.	.725	74.12	19.38	93.50	
Trim, casing, painted	34.000	L.F.	1.380	29.58	39.78	69.36	
Butt hinges, chrome, 3-1/2'' x 3-1/2''	1.500	Pr.		23.85		23.85	
Lockset, passage	1.000	Ea.	.500	12.45	14.70	27.15	
Paint, door & frame, primer & 2 coats	2.000	Face	2.465	9.72	64.80	74.52	
TOTAL			5.959	205.22	162.79	368.01	
RAISED PANEL, SOLID, PINE DOOR							
Door, pine, raised panel, 2'-8'' wide x 6'-8'' high	1.000	Ea.	.889	149.00	24.13	173.13	
Frame, pine, 4-5/8'' jamb	17.000	L.F.	.725	74.12	19.38	93.50	
Trim, casing, painted	34.000	L.F.	1.380	29.58	39.78	69.36	
Butt hinges, bronze, 3-1/2'' x 3-1/2''	1.500	Pr.		27.15		27.15	
Lockset, passage	1.000	Ea.	.500	12.45	14.70	27.15	
Paint, door & frame, primer & 2 coats	2.000	Face	3.067	13.68	81.10	94.78	
TOTAL			6.561	305.98	179.09	485.07	

The costs in these systems are based on a cost per each door.

Interior Door System Components

Component Description	QUAN.	UNIT	LABOR-HOURS	MAT.	INST.	TOTAL
Doors						
Door, hollow core, lauan 1-3/8'' thick, 6'-8'' high x 1'-6'' wide	1.000	Ea.	.842	37.50	22.50	60.00
2'-6'' wide	1.000	Ea.	.889	43.50	24.00	67.50
3'-0'' wide	1.000	Ea.	.941	47.50	25.00	72.50
Birch 1-3/8'' thick, 6'-8'' high x 1'-6'' wide	1.000	Ea.	.842	42.00	22.50	64.50
2'-6'' wide	1.000	Ea.	.889	53.00	24.00	77.00
3'-0'' wide	1.000	Ea.	.941	58.00	25.00	83.00
Louvered pine 1-3/8'' thick, 6'-8'' high x 1'-6'' wide	1.000	Ea.	.842	100.00	22.50	122.50
2'-0'' wide	1.000	Ea.	.889	124.00	24.00	148.00
2'-6'' wide	1.000	Ea.	.889	140.00	24.00	164.00
2'-8'' wide	1.000	Ea.	.889	146.00	24.00	170.00
3'-0'' wide	1.000	Ea.	.941	156.00	25.00	181.00
Paneled pine 1-3/8'' thick, 6'-8'' high x 1'-6'' wide	1.000	Ea.	.842	107.00	22.50	129.50
2'-6'' wide	1.000	Ea.	.889	138.00	24.00	162.00
3'-0'' wide	1.000	Ea.	.941	164.00	25.00	189.00

Closet Door Systems

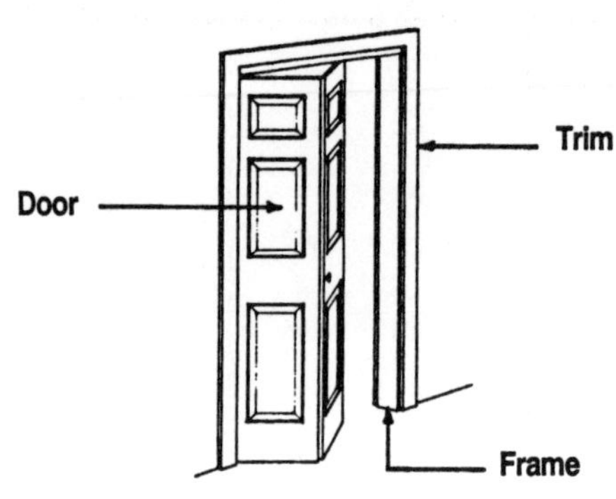

System Description	QUAN.	UNIT	LABOR-HOURS	COST EACH			EA. EXTENSIONS
				MAT.	INST.	TOTAL	
BI-PASSING, FLUSH, LAUAN, HOLLOW CORE, 4'-0'' X 6'-8''							
Door, flush, lauan, hollow core, 4'-0'' x 6'-8'' opening	1.000	Ea.	1.333	114.00	35.69	149.69	
Frame, pine, 4-5/8'' jamb	18.000	L.F.	.768	78.48	20.52	99.00	
Trim, both sides, casing, painted	36.000	L.F.	1.849	29.88	52.20	82.08	
Paint, door & frame, primer & 2 coats	2.000	Face	2.465	9.72	64.80	74.52	
TOTAL			6.415	232.08	173.21	405.29	
BI-PASSING, PINE, PANELED, 4'-0'' X 6'-8''							
Door, pine, paneled, 4'-0'' x 6'-8'' opening	1.000	Ea.	1.333	288.00	35.69	323.69	
Frame, pine, 4-5/8'' jamb	18.000	L.F.	.768	78.48	20.52	99.00	
Trim, both sides, painted	36.000	L.F.	1.849	29.88	52.20	82.08	
Paint, door & frame, primer & 2 coats	2.000	Face	3.067	13.68	81.10	94.78	
TOTAL			7.017	410.04	189.51	599.55	
BI-PASSING, FLUSH, BIRCH, HOLLOW CORE, 6'-0'' X 6'-8''							
Door, flush, birch, hollow core, 6'-0'' x 6'-8'' opening	1.000	Ea.	1.600	151.00	43.03	194.03	
Frame, pine, 4-5/8'' jamb	19.000	L.F.	.811	82.84	21.66	104.50	
Trim, both sides, casing, painted	38.000	L.F.	1.952	31.54	55.10	86.64	
Paint, door & frame, primer & 2 coats	2.000	Face	3.081	12.15	81.00	93.15	
TOTAL			7.444	277.53	200.79	478.32	
BI-PASSING, PINE, PANELED, 6'-0'' X 6'-8''							
Door, pine, paneled, 6'-0'' x 6'-8'' opening	1.000	Ea.	1.600	370.00	43.03	413.03	
Frame, pine, 4-5/8'' jamb	19.000	L.F.	.811	82.84	21.66	104.50	
Trim, both sides, painted	38.000	L.F.	1.952	31.54	55.10	86.64	
Paint, door & frame, primer & 2 coats	2.000	Face	3.833	17.10	101.38	118.48	
TOTAL			8.196	501.48	221.17	722.65	
BI-FOLD, PINE, PANELED, 3'-0'' X 6'-8''							
Door, pine, paneled, 3'-0'' x 6'-8'' opening	1.000	Ea.	1.231	84.50	33.06	117.56	
Frame, pine, 4-5/8'' jamb	17.000	L.F.	.725	74.12	19.38	93.50	
Trim, both sides, casing, painted	34.000	L.F.	1.746	28.22	49.30	77.52	
Paint, door & frame, primer & 2 coats	2.000	Face	3.067	13.68	81.10	94.78	
TOTAL			6.769	200.52	182.84	383.36	

Closet Door Systems

System Description	QUAN.	UNIT	LABOR-HOURS	COST EACH			EA. EXTENSIONS
				MAT.	INST.	TOTAL	
BI-FOLD, PINE, LOUVERED, 3'-0'' X 6'-8''							
Door, pine, louvered, 3'-0'' x 6'-8'' opening	1.000	Ea.	1.231	84.50	33.06	117.56	
Frame, pine, 4-5/8'' jamb	17.000	L.F.	.725	74.12	19.38	93.50	
Trim, both sides, painted	34.000	L.F.	1.746	28.22	49.30	77.52	
Paint, door & frame, primer & 2 coats	2.000	Face	2.300	10.26	60.83	71.09	
TOTAL			6.002	197.10	162.57	359.67	
BI-FOLD, PINE, LOUVERED, 6'-0'' X 6'-8''							
Door, pine, louvered, 6'-0'' x 6'-8'' opening	1.000	Ea.	1.600	165.00	43.03	208.03	
Frame, pine, 4-5/8'' jamb	19.000	L.F.	.811	82.84	21.66	104.50	
Trim, both sides, casing, painted	38.000	L.F.	1.952	31.54	55.10	86.64	
Paint, door & frame, primer & 2 coats	2.000	Face	3.833	17.10	101.38	118.48	
TOTAL			8.196	296.48	221.17	517.65	
BI-FOLD, FLUSH, BIRCH, 6'-0'' X 6'-8''							
Door, flush, birch, hollow core, 6'-0'' x 6'-8'' opng.	1.000	Ea.	1.600	105.00	43.03	148.03	
Frame, pine, 4-5/8'' jamb	19.000	L.F.	.811	82.84	21.66	104.50	
Trim, both sides, painted	38.000	L.F.	1.952	31.54	55.10	86.64	
Paint, door & frame, primer & 2 coats	2.000	Face	3.081	12.15	81.00	93.15	
TOTAL			7.444	231.53	200.79	432.32	

The costs in this system are based on a cost per each door.

Closet Door System Components

Component Description	QUAN.	UNIT	LABOR-HOURS	COST EACH		
				MAT.	INST.	TOTAL
Doors						
Doors, bi-passing, pine, louvered, 4'-0'' x 6'-8'' opening	1.000	Ea.	1.333	278.00	35.50	313.50
6'-0'' x 6'-8'' opening	1.000	Ea.	1.600	350.00	43.00	393.00
Flush, birch, hollow core, 4'-0'' x 6'-8'' opening	1.000	Ea.	1.333	128.00	35.50	163.50
6'-0'' x 6'-8'' opening	1.000	Ea.	1.600	151.00	43.00	194.00
Flush, lauan, hollow core, 4'-0'' x 6'-8'' opening	1.000	Ea.	1.333	114.00	35.50	149.50
6'-0'' x 6'-8'' opening	1.000	Ea.	1.600	130.00	43.00	173.00
Bi-fold, pine, louvered, 3'-0'' x 6'-8'' opening	1.000	Ea.	1.231	84.50	33.00	117.50
6'-0'' x 6'-8'' opening	1.000	Ea.	1.600	165.00	43.00	208.00
Flush, birch, hollow core, 3'-0'' x 6'-8'' opening	1.000	Ea.	1.231	53.00	33.00	86.00
6'-0'' x 6'-8'' opening	1.000	Ea.	1.600	105.00	43.00	148.00
Flush, lauan, hollow core, 3'-0'' x 6'8'' opening	1.000	Ea.	1.231	141.00	33.00	174.00
6'-0'' x 6'-8'' opening	1.000	Ea.	1.600	277.00	43.00	320.00
Frames						
Frame pine, 3'-0'' door, 3-5/8'' deep	17.000	L.F.	.725	69.50	19.40	88.90
5-5/8'' deep	17.000	L.F.	.725	84.00	19.40	103.40
4'-0'' door, 3-5/8'' deep	18.000	L.F.	.768	73.50	20.50	94.00
5-5/8'' deep	18.000	L.F.	.768	88.50	20.50	109.00
6'-0'' door, 3-5/8'' deep	19.000	L.F.	.811	77.50	21.50	99.00
5-5/8'' deep	19.000	L.F.	.811	93.50	21.50	115.00

Construction Information

Provided for your use are illustrations together with the common terms used for the various components of framing and rough carpentry. The reason for this is that terms used are not universal. What someone in one part of the country might call a bottom plate, someone else might call a sole plate.

In this section are the common names used for various exterior wood framing systems, siding systems and roof framing systems. Illustrations and terminology are also shown for various interior systems such as partition framing systems, ceiling systems, and door and window systems.

Wood Framing Nomenclature

These graphics illustrate the common terms used for the various components of wood framing. Whether for residential (shown here) or commercial, the terminology is the same.

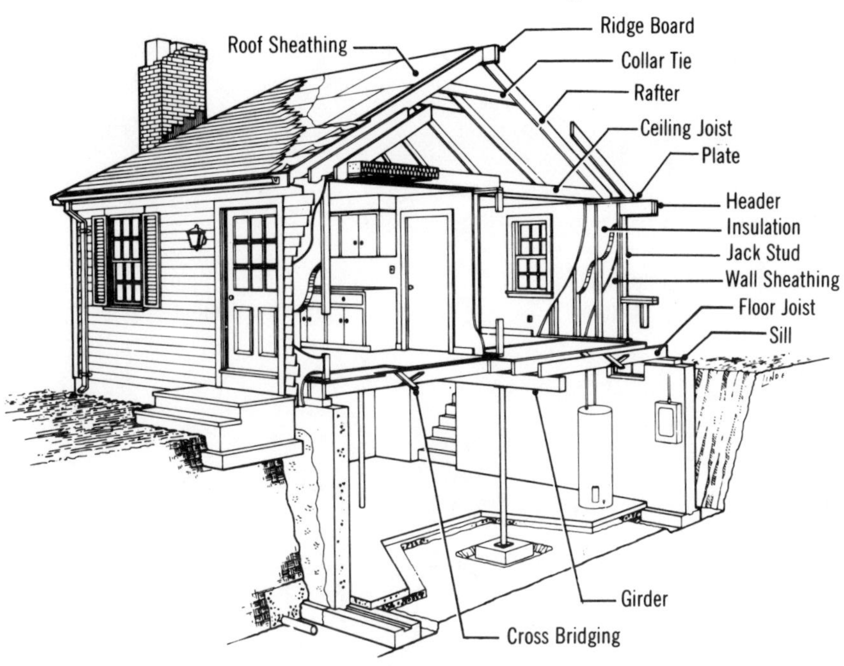

Wood Framing System

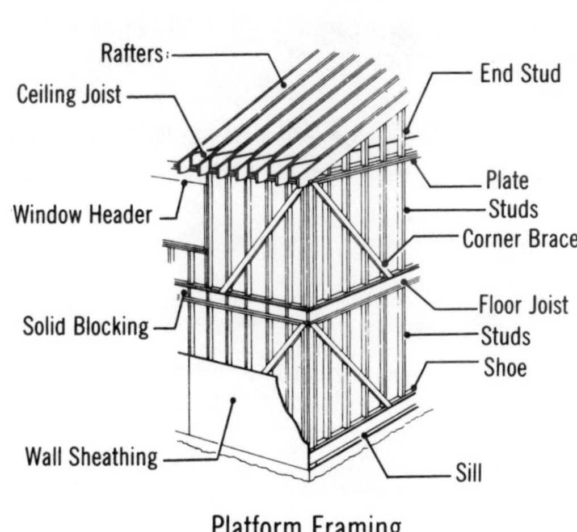

Platform Framing

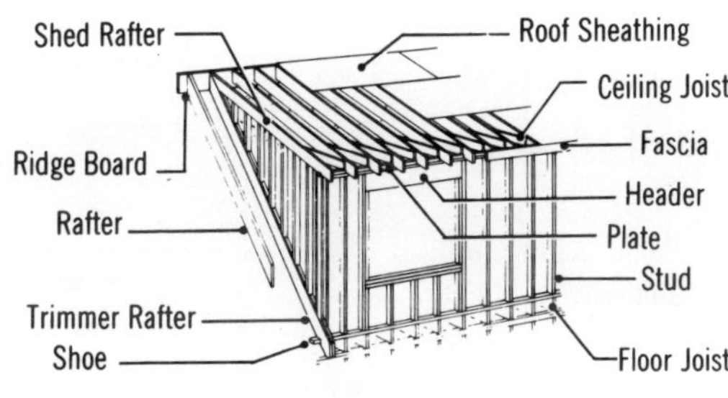

Shed Dormer Framing

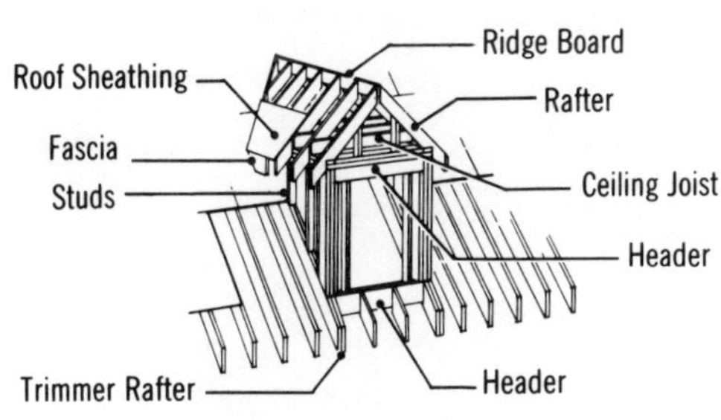

Gable Dormer Framing

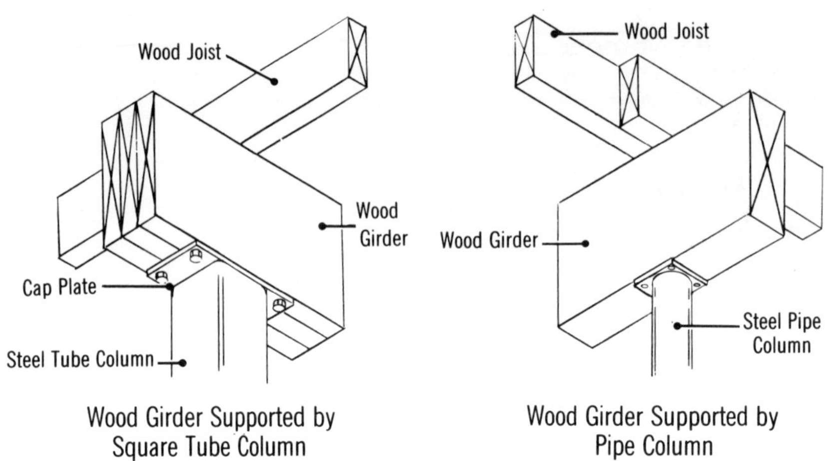

Wood Girder Supported by
Square Tube Column

Wood Girder Supported by
Pipe Column

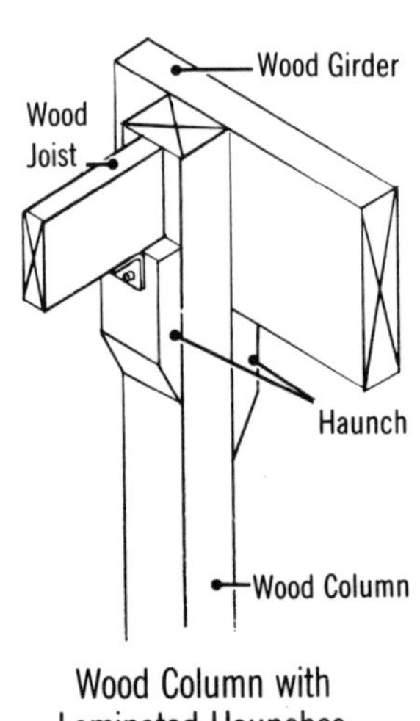

Wood Column with
Laminated Haunches

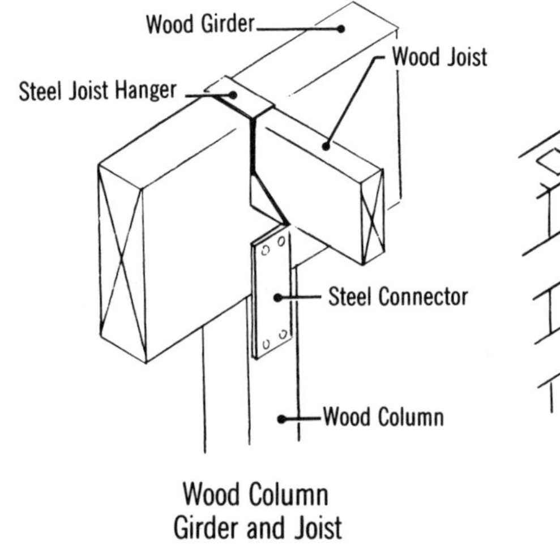

Wood Column
Girder and Joist

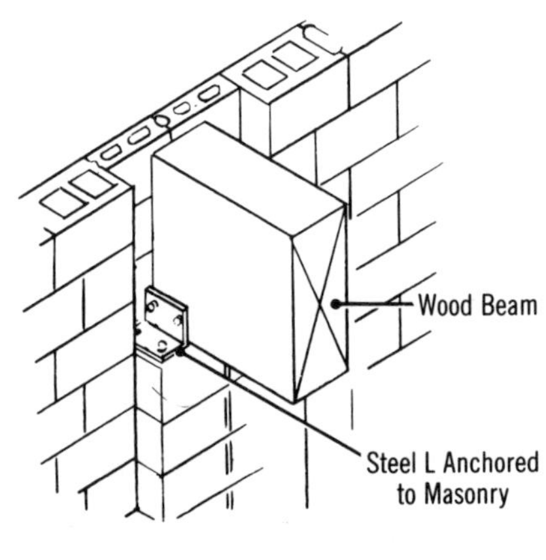

Wood Girder Supported
by Masonry Wall

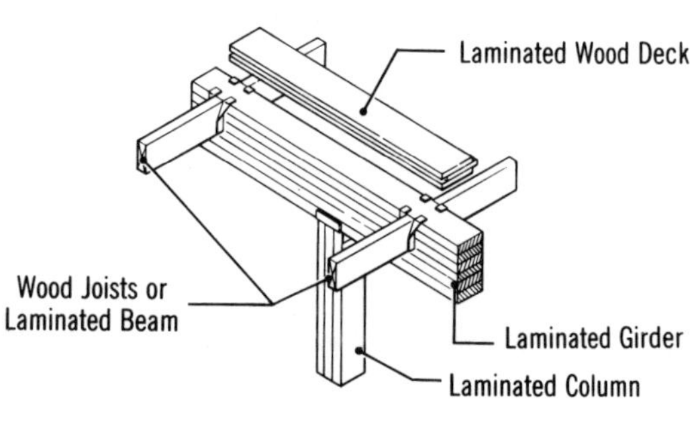

Laminated Wood Floor Beams

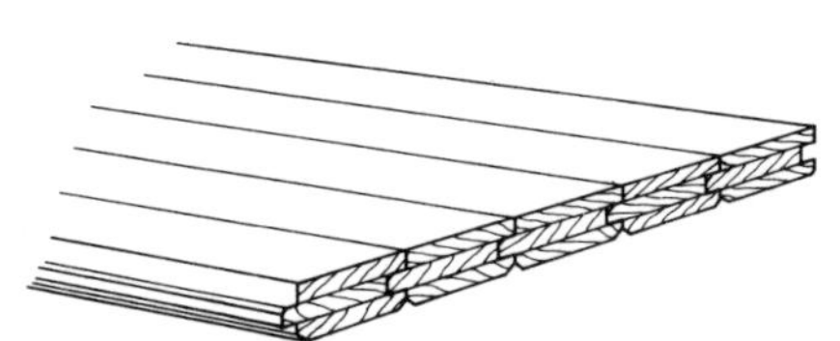

Laminated Wood Deck

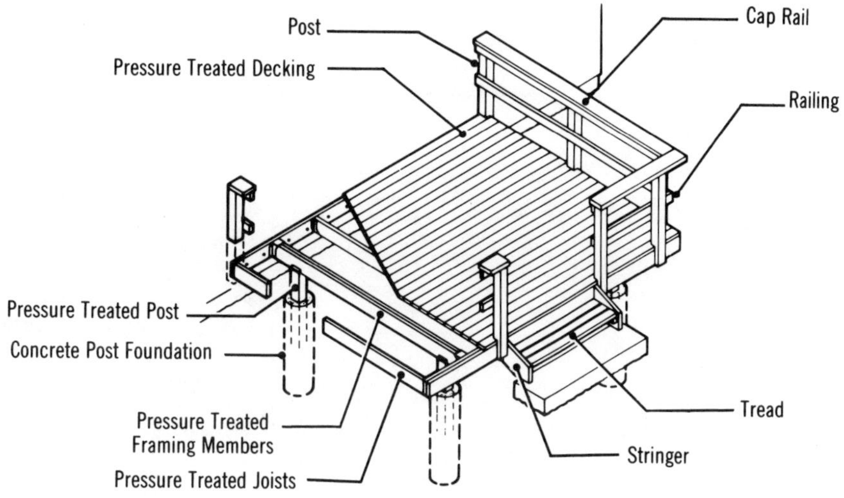

Wood Deck Construction

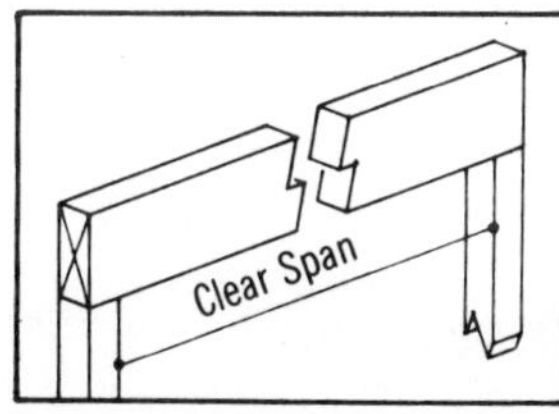

Single Beam

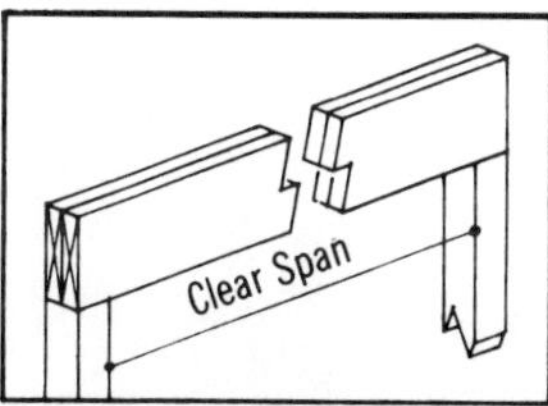

Double Beam

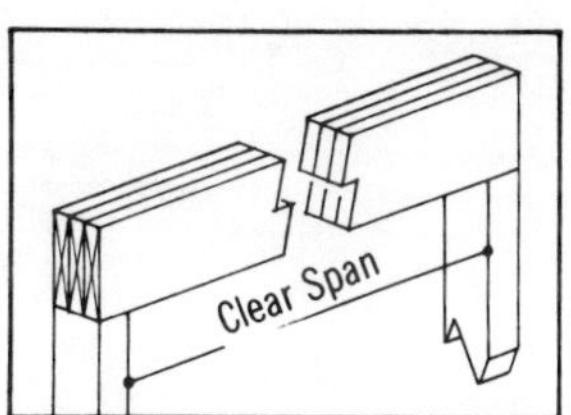

Triple Beam

Wood Beams and Columns

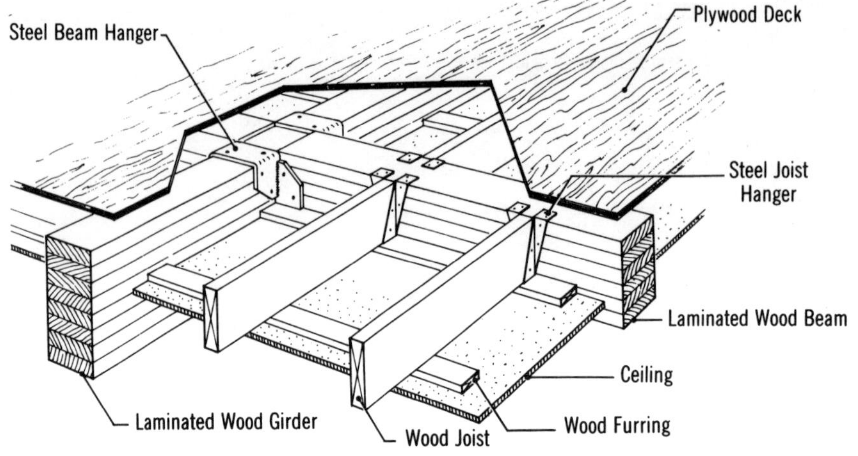

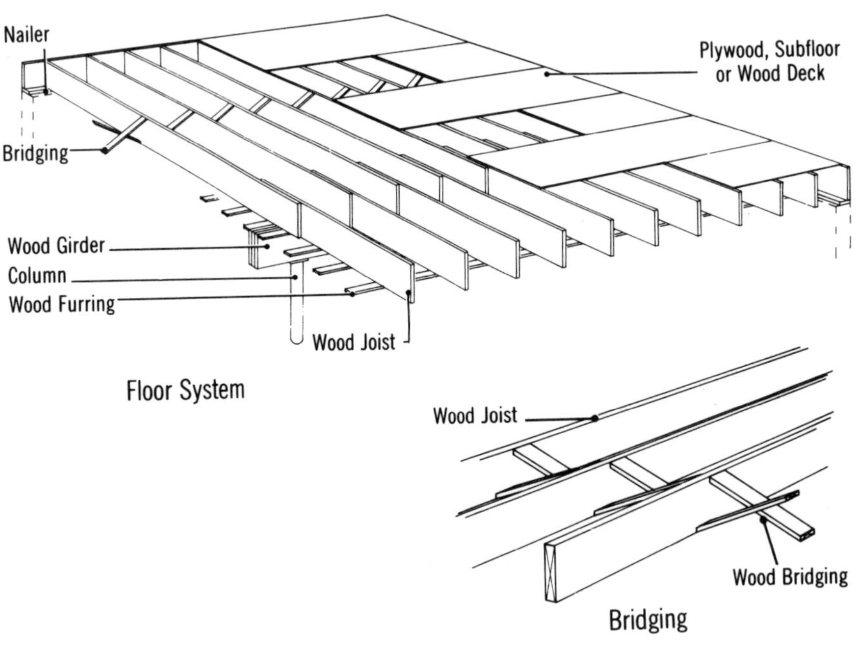

Floor System

Bridging

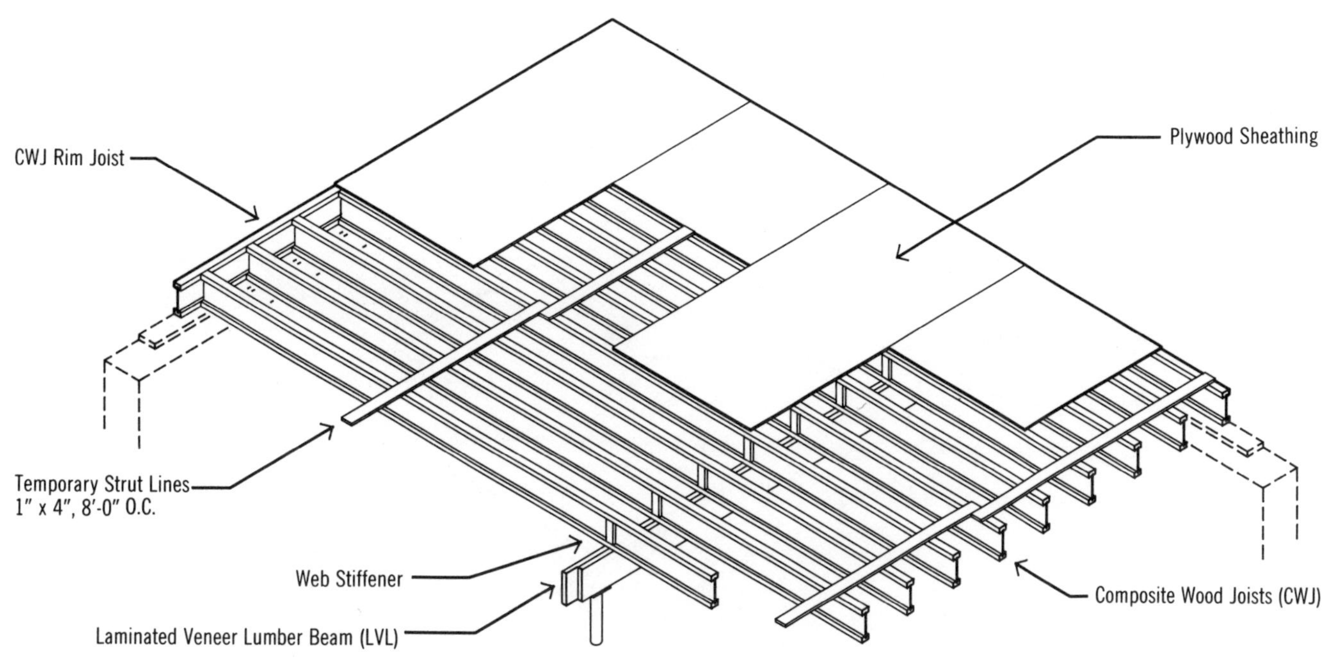

Composite Wood Joists

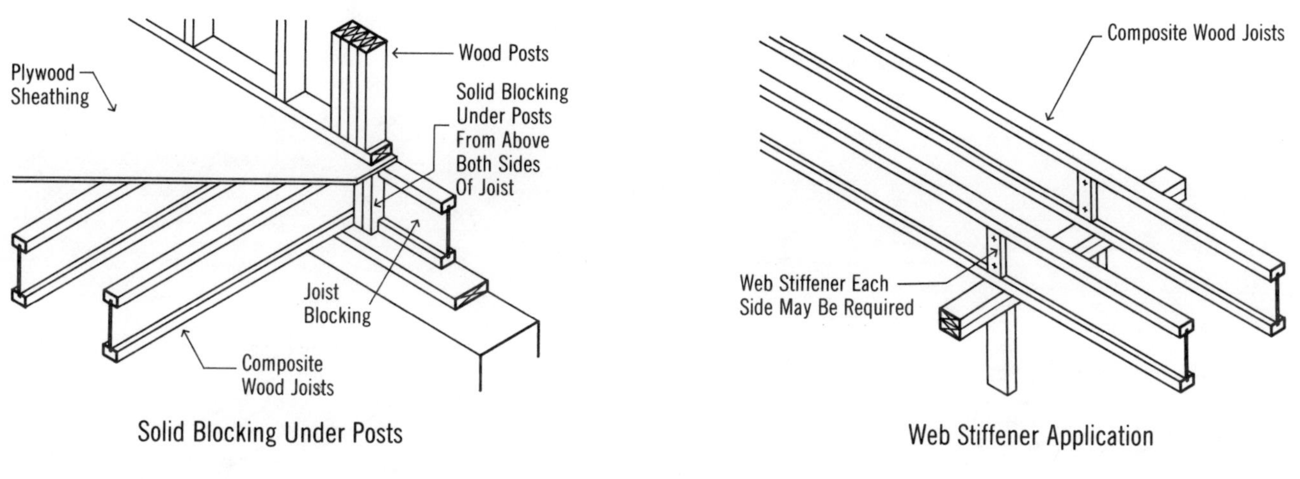

Solid Blocking Under Posts

Web Stiffener Application

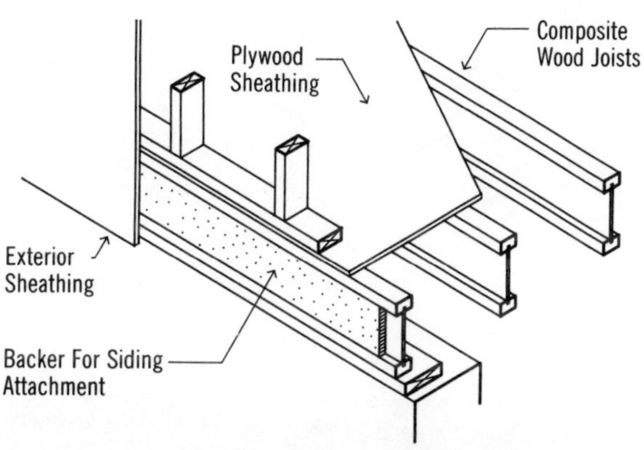

Rim Joist Backer Board Attachment

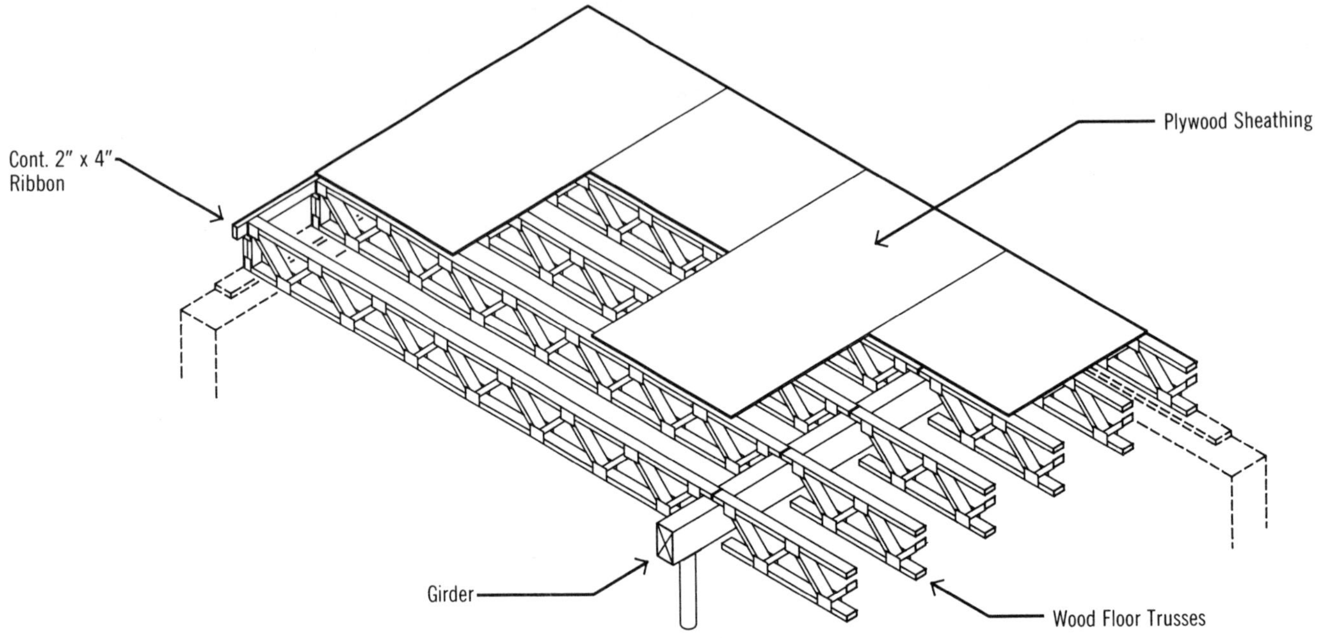

Wood Floor Trusses

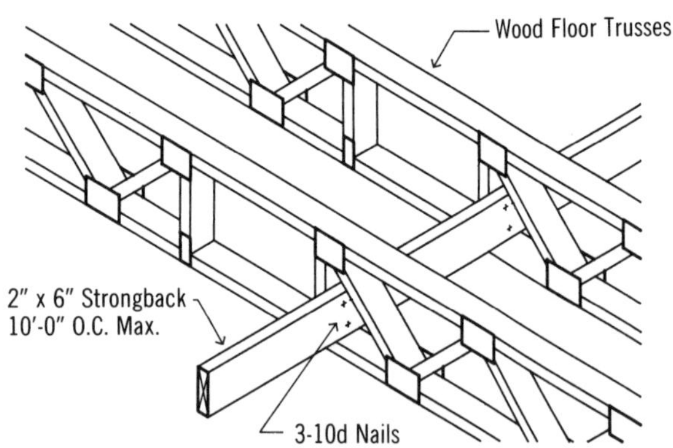

Application of Strongback

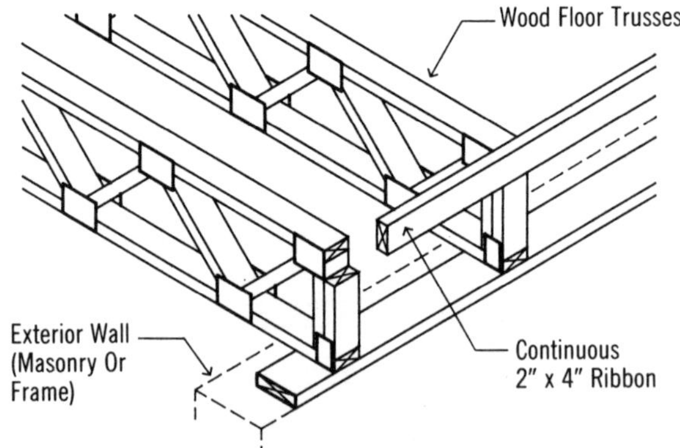

Bottom Chord Exterior Wall Bearing

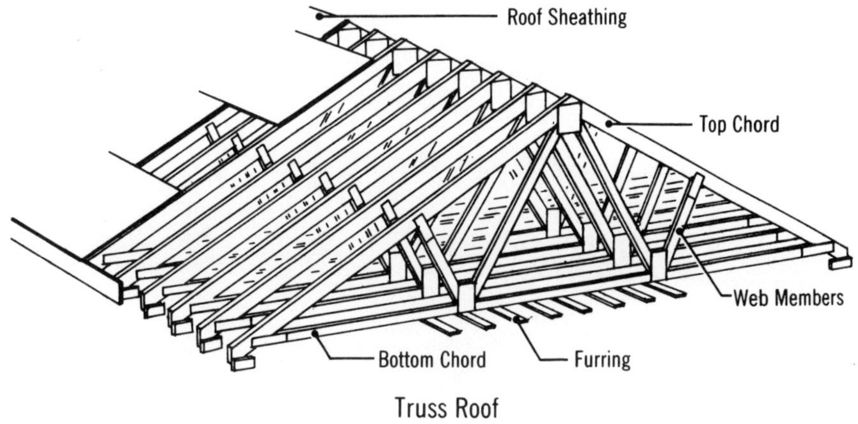

Truss Roof

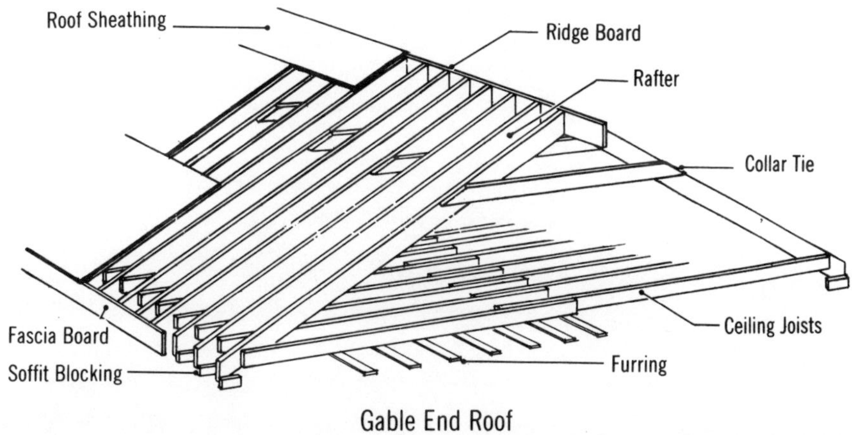

Gable End Roof

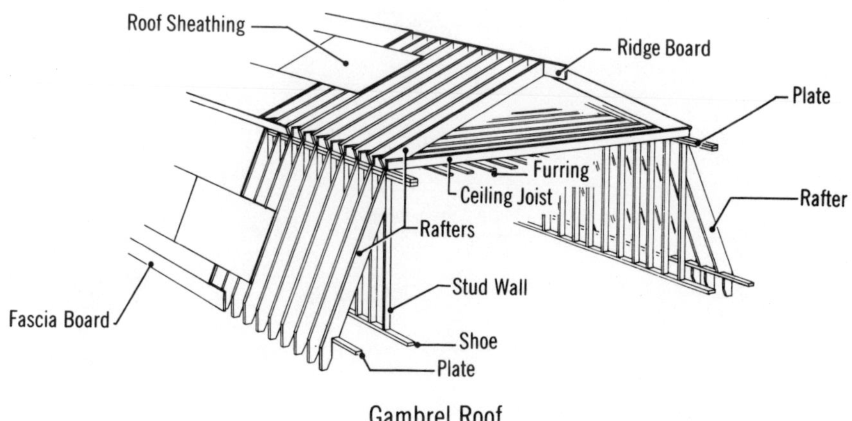

Gambrel Roof

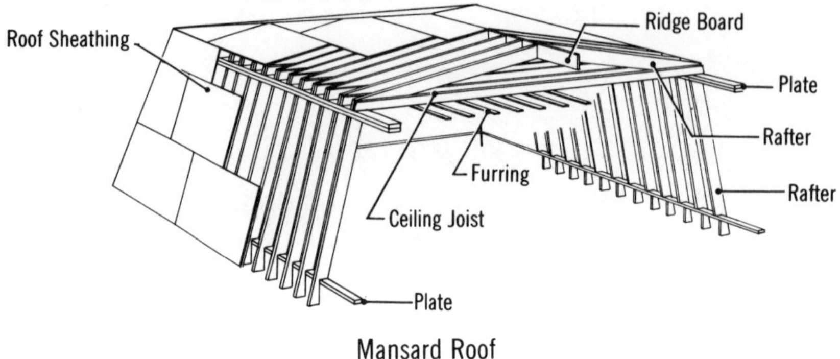

Mansard Roof

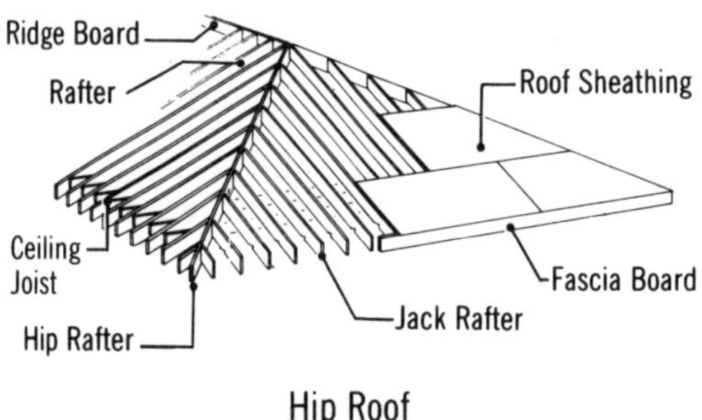

Hip Roof

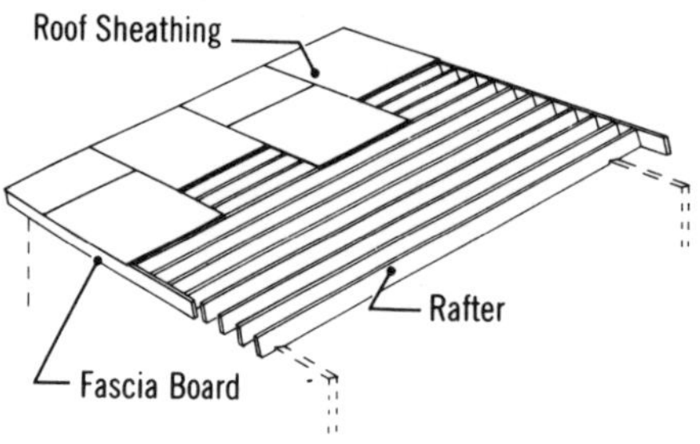

Shed Roof

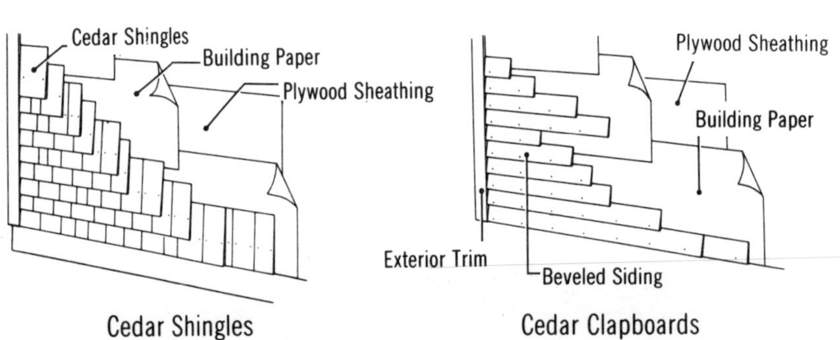

Cedar Shingles   Cedar Clapboards

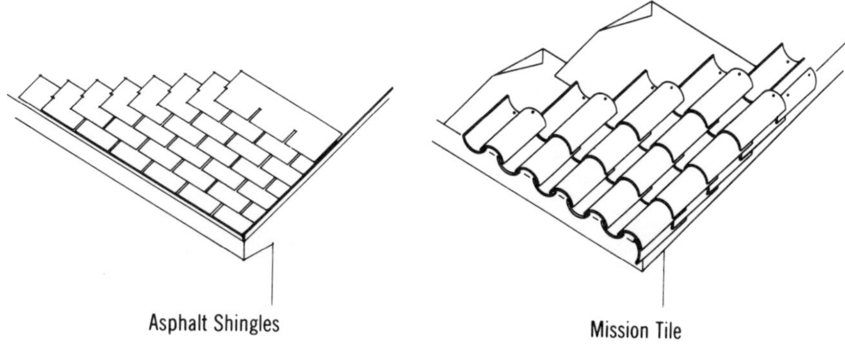

Asphalt Shingles

Mission Tile

Water Shed Roof Coverings

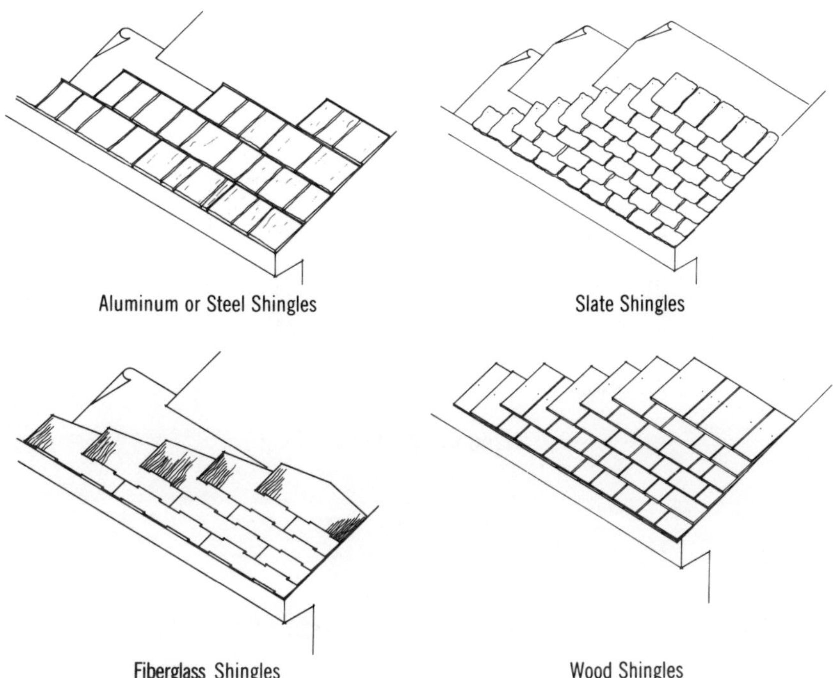

Aluminum or Steel Shingles

Slate Shingles

Fiberglass Shingles

Wood Shingles

Water Shed Roof Coverings

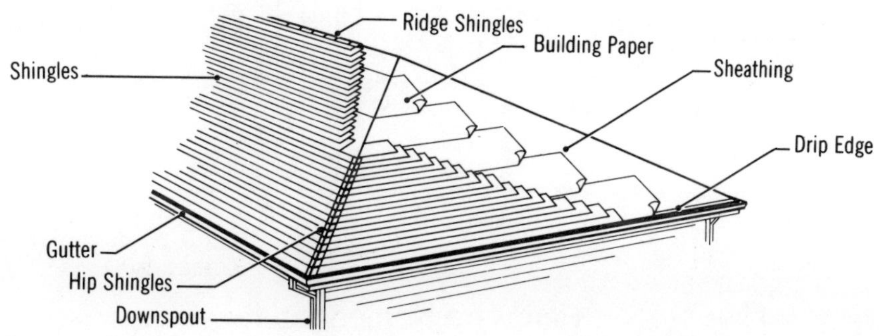

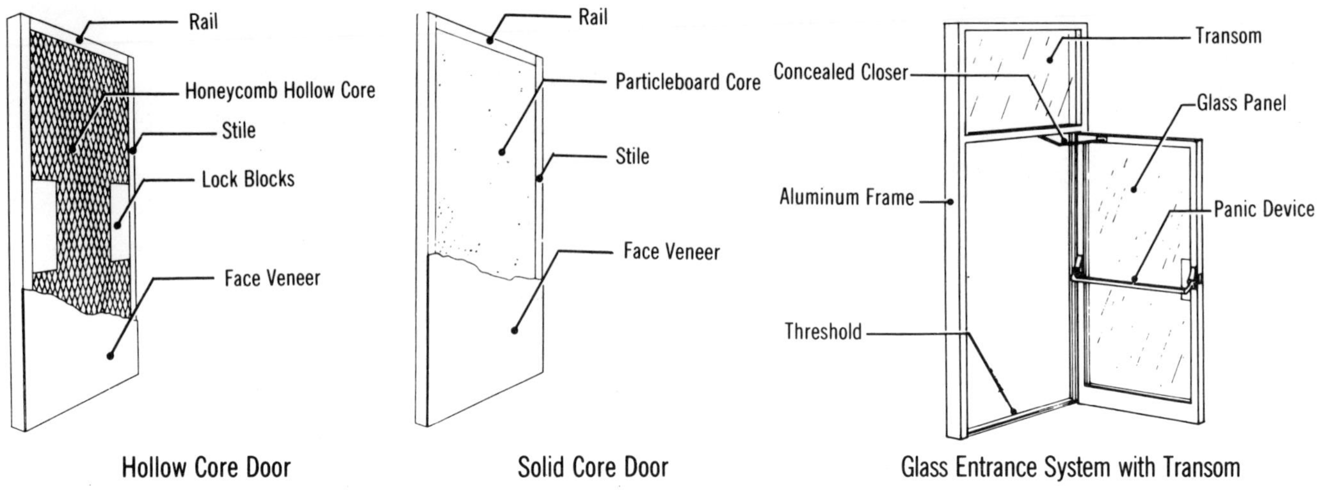

Hollow Core Door

Solid Core Door

Glass Entrance System with Transom

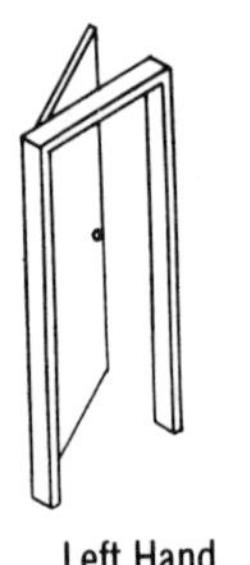

Left Hand

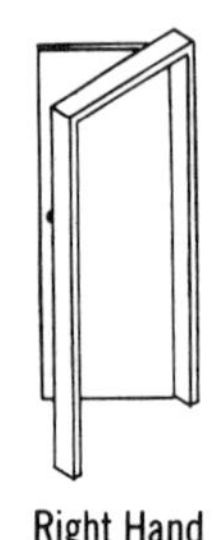

Right Hand

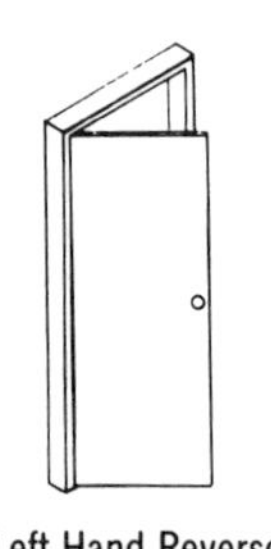

Left Hand Reverse

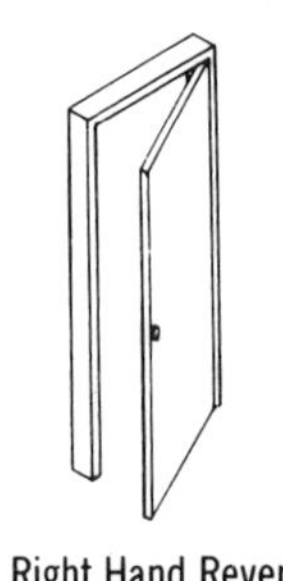

Right Hand Reverse

Hand Designations

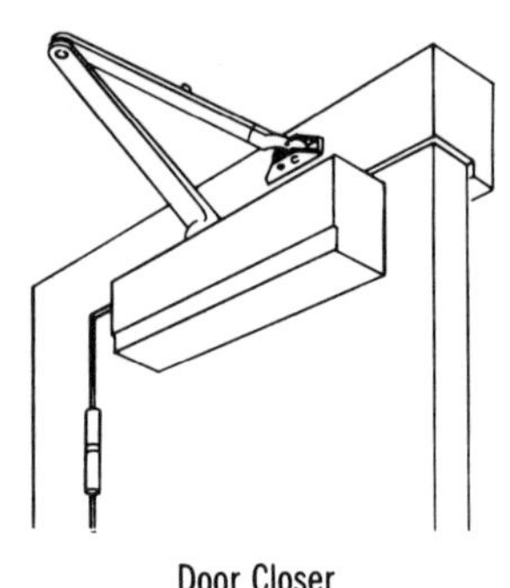

Door Closer

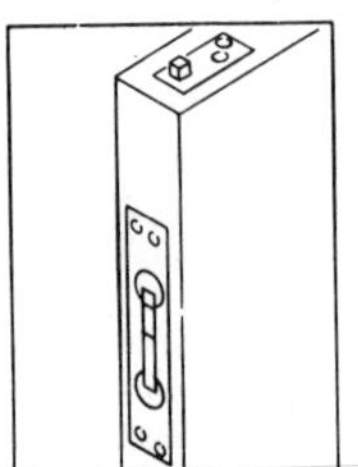

Flush Bolt, Concealed

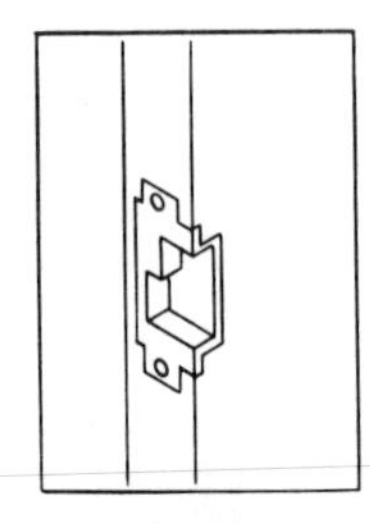

Open Back Strike

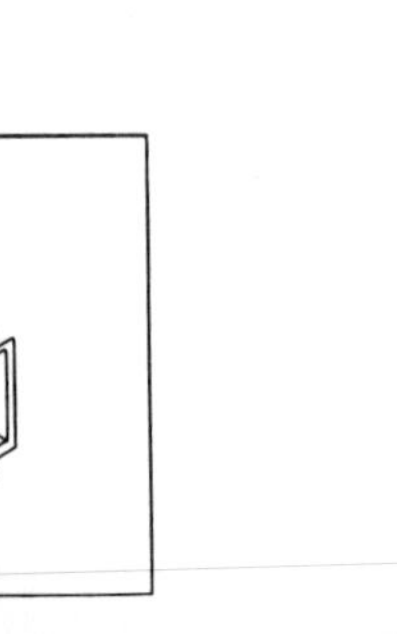

Locksets

Hardware Nomenclature

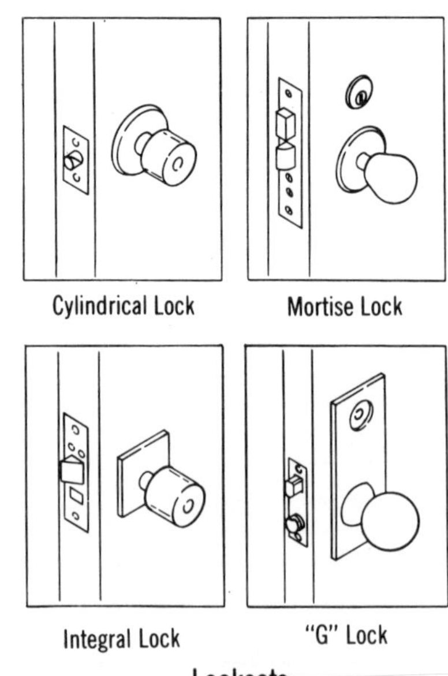

Rim Mounted Panic Bar

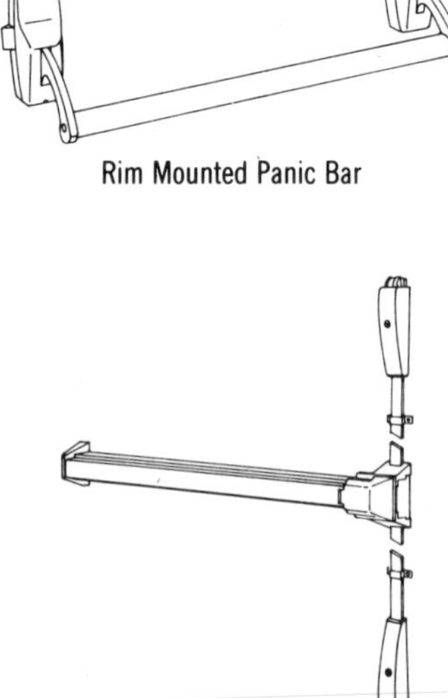

Touch Bar and Vertical Rod

Panic Devices

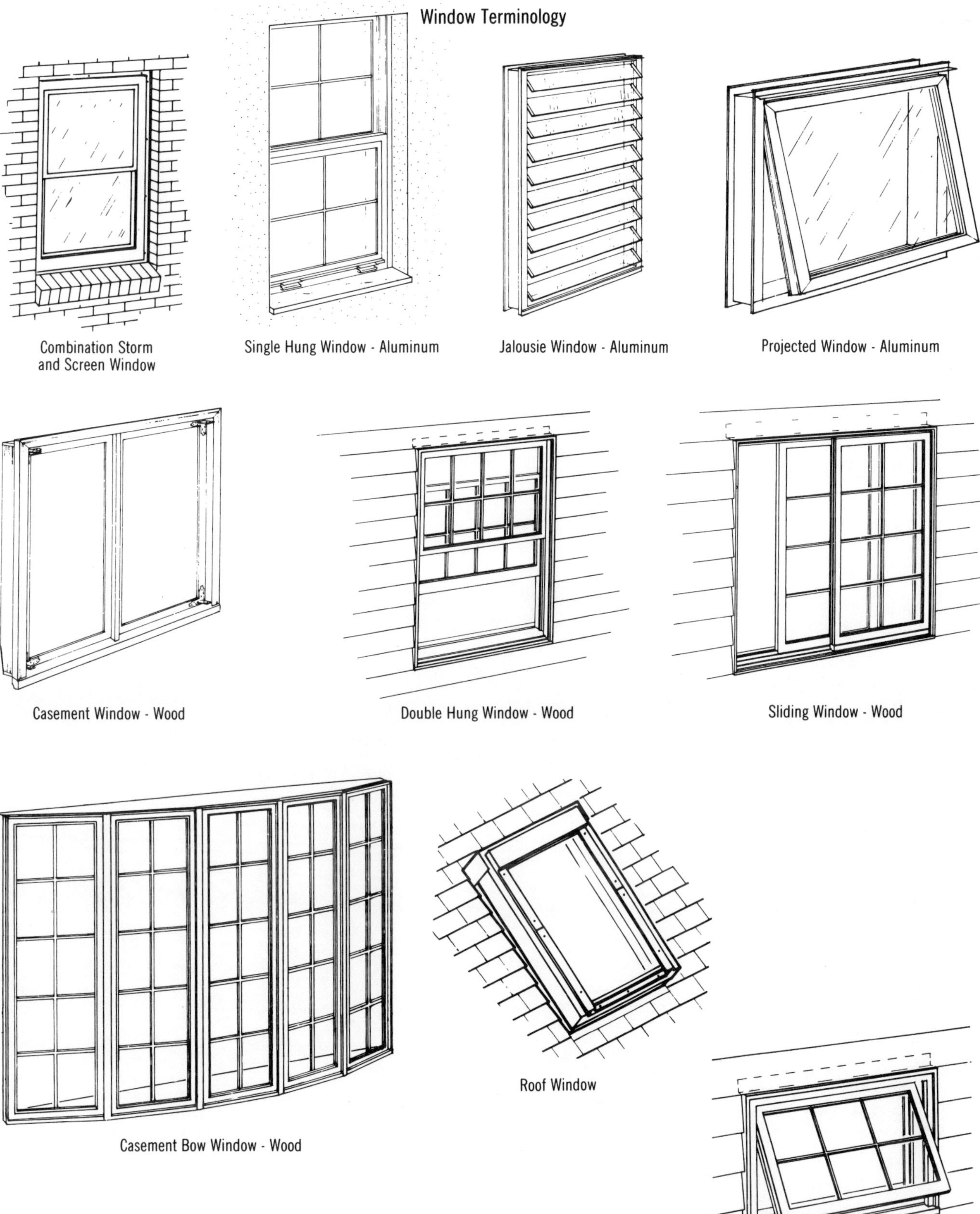

Window Terminology
Combination Storm and Screen Window
Single Hung Window - Aluminum
Jalousie Window - Aluminum
Projected Window - Aluminum
Casement Window - Wood
Double Hung Window - Wood
Sliding Window - Wood
Casement Bow Window - Wood
Roof Window
Awning Window - Wood

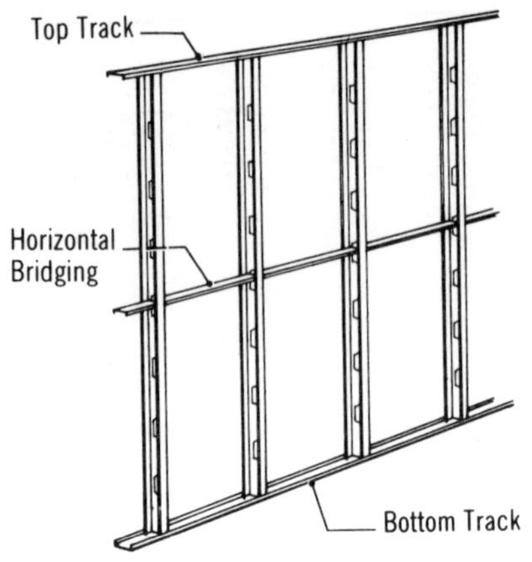

Load Bearing Steel Studs

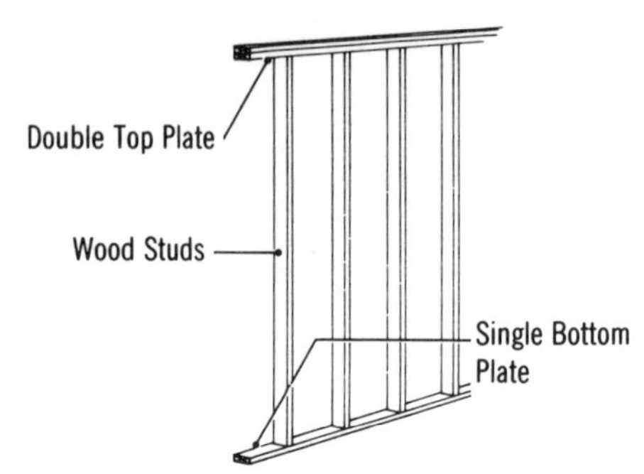

Wood Stud Partition, No Blocking

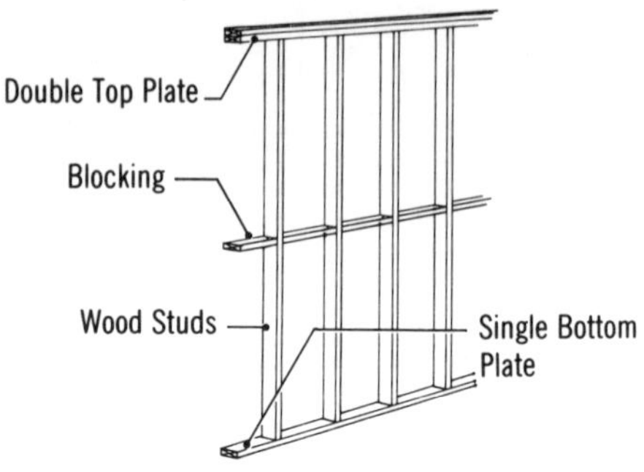

Wood Stud Partition with Blocking

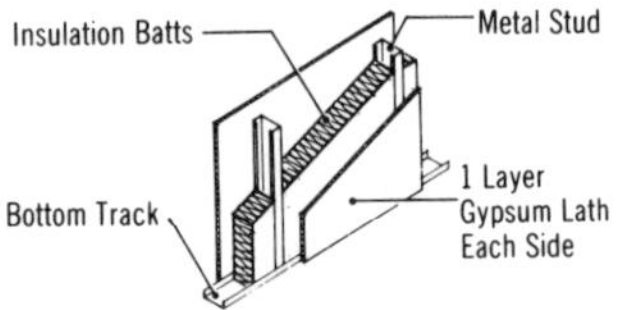

Fiberglass Batt Insulation

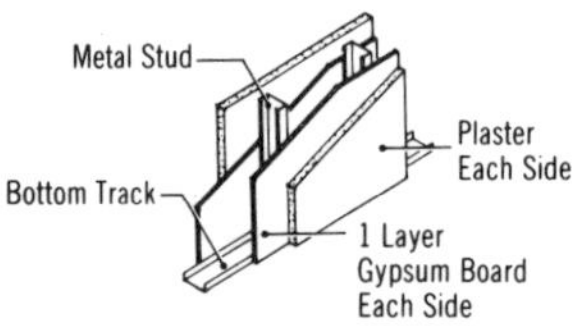

Plaster on Gypsum Lath

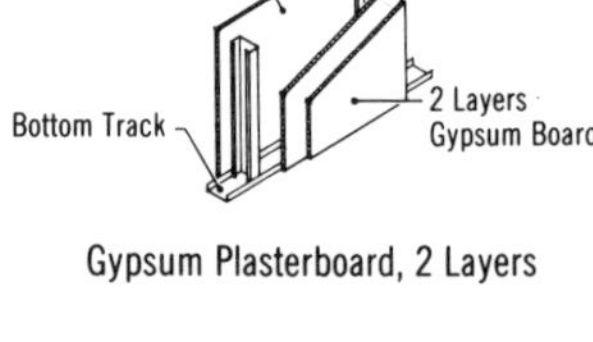

Gypsum Plasterboard, 2 Layers

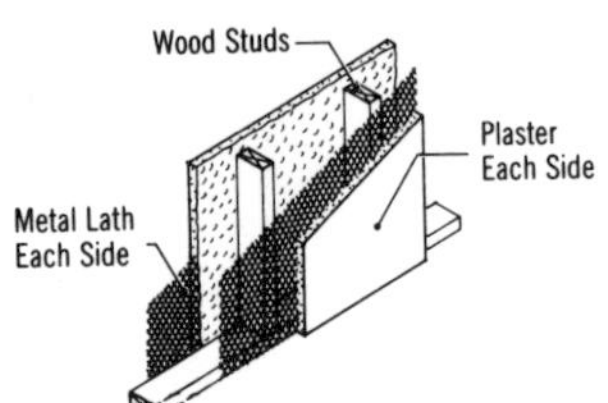

Plaster on Metal Lath

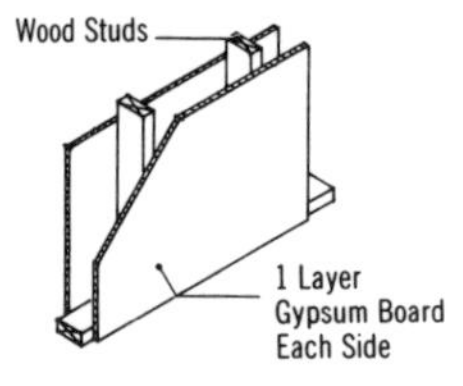

Gypsum Plasterboard

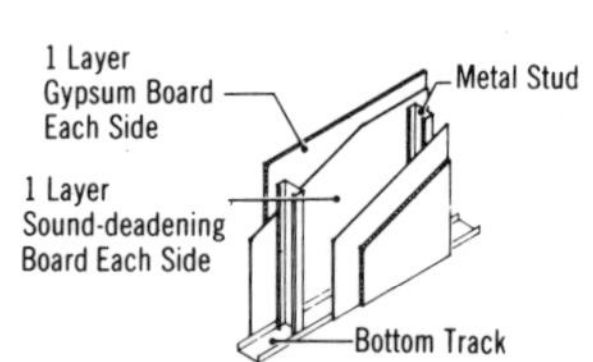

Plaster on Gypsum Lath

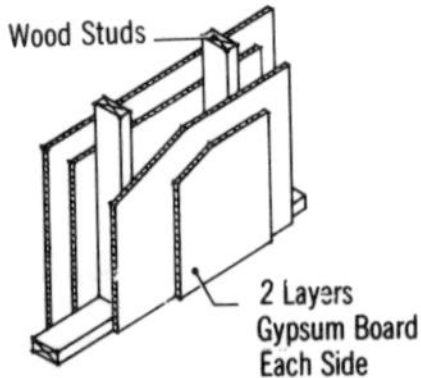

Plaster on Metal Lath

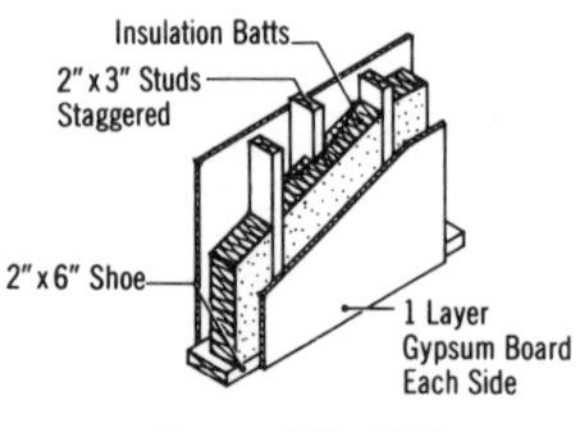

Gypsum Plasterboard

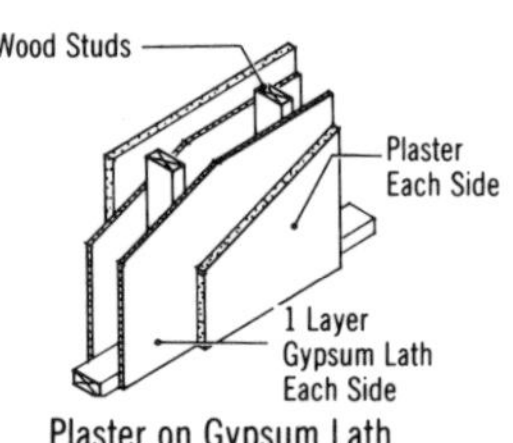

Sound-deadening Board

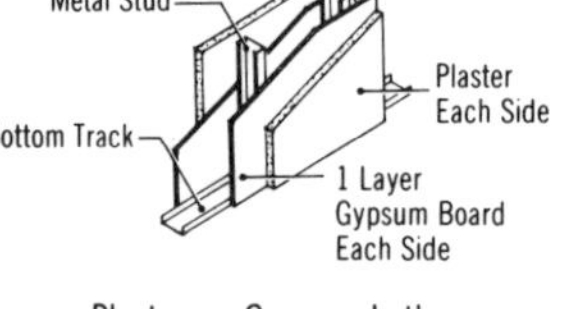

Gypsum Plasterboard, 2 Layers

Staggered Stud Wall

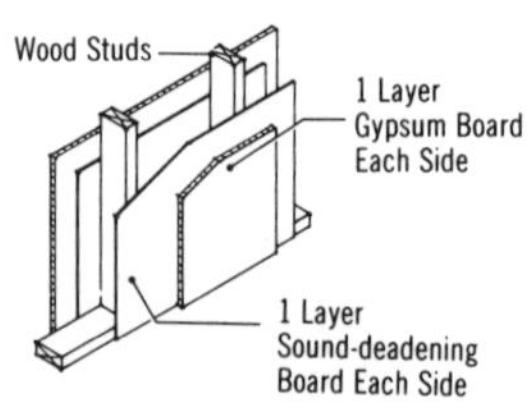

Sound-deadening Board

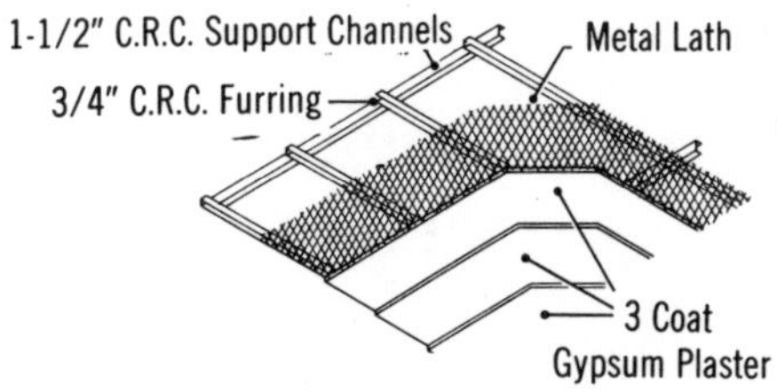

Plaster on Metal Lath

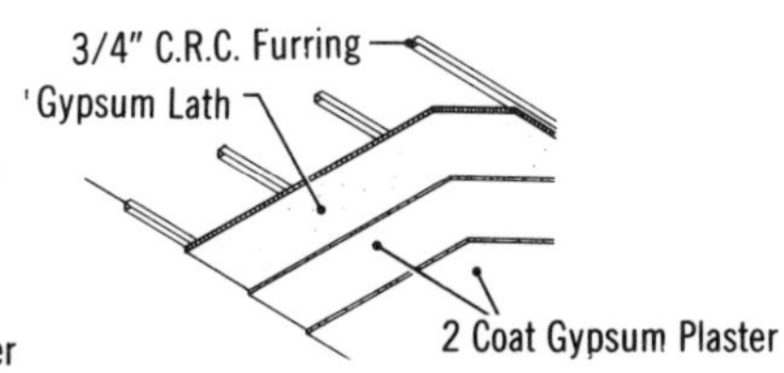

Plaster on Metal Furring

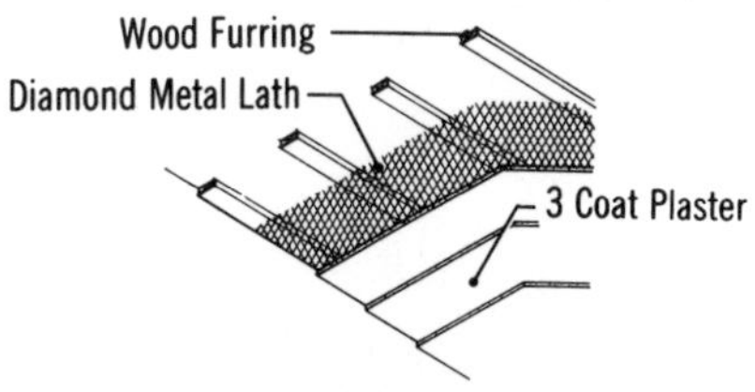

**Plaster on Metal Lath
and Wood Furring**

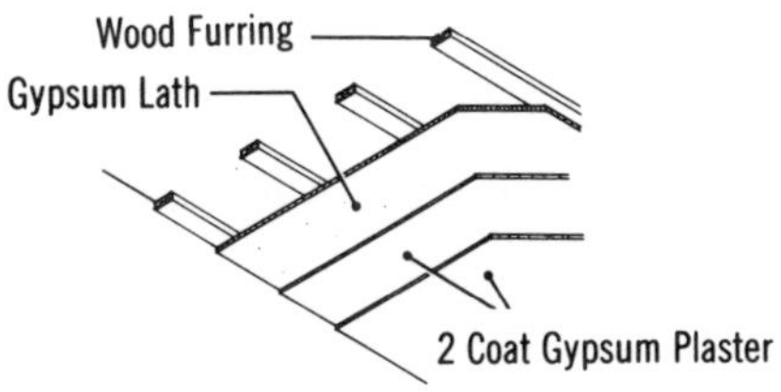

Plaster on Wood Furring

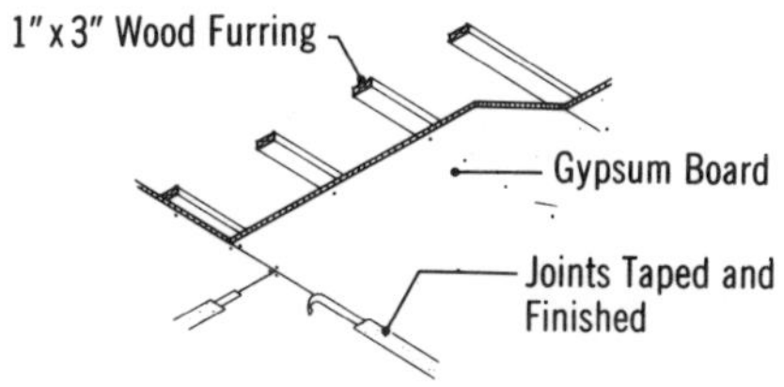

Gypsum Board on 1" x 3" Wood Furring

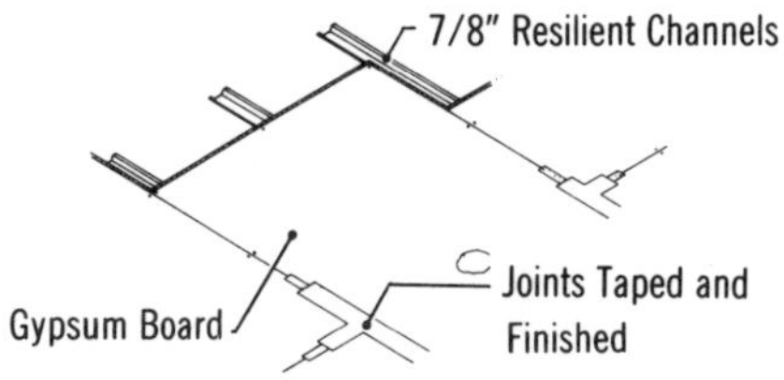

**Gypsum Board
on 7/8" Resilient Channel Furring**

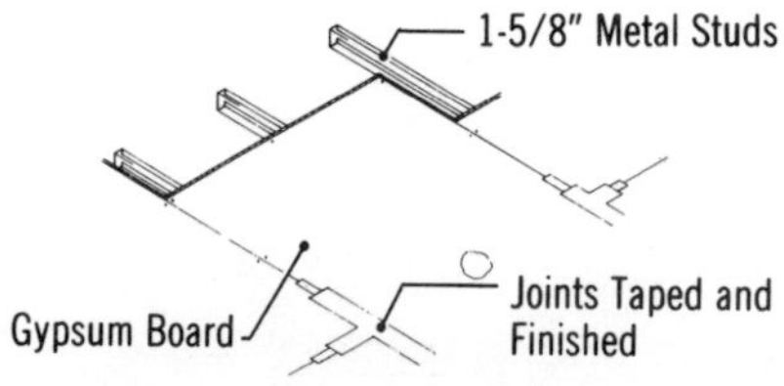

**Gypsum Board
on 1-5/8" Metal Stud Furring**

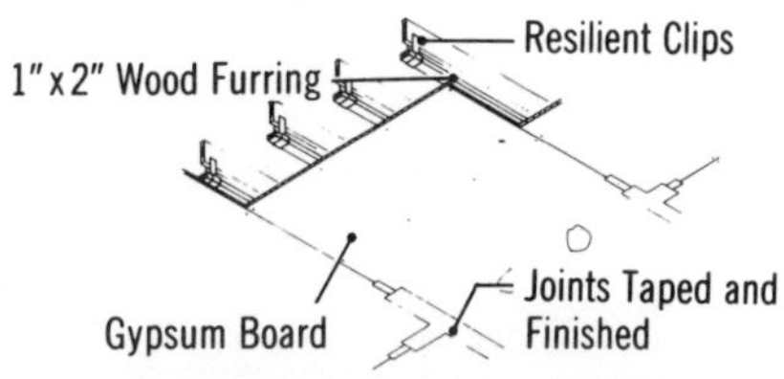

**Gypsum Board on 1" x 2",
Suspended, with Resilient Clips**

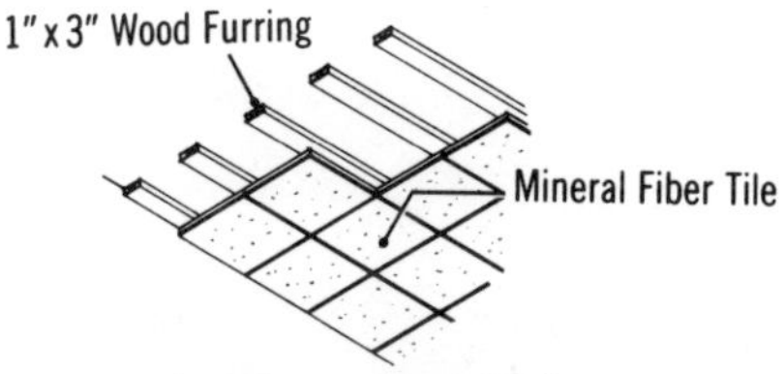

**Acoustical Mineral Fiber Tile
on 1" x 3" Wood Furring**

**Acoustical Mineral Fiber Tile
on Gypsum Board**

Mineral Fiber Tile Applied with Adhesive

Section Three
Business Information

Total project costs include more than costs for the material, labor, and equipment necessary for the completion of your project. These cost categories include project overhead, general conditions, profit, fees, permits, main office expenses, insurance, tools, taxes, and clean-up to name a few.

This business information section not only provides you with insight into these additional cost factors, but also includes some solid information concerning how to select the most desirable projects for your company to bid, how to manage your existing projects, and some basics concerning scheduling your projects for better profitability.

Business Information

An estimate can be divided into two main groups of costs — direct and indirect. By definition direct costs are those costs that can be directly linked to the project, those costs without which the project could not be completed. Included in this group are material, labor, equipment, and subcontractor costs. These costs are sometimes labeled "bare" or "unburdened" costs.

Direct costs of a construction project include material, labor, equipment (used on the job site), subcontractors, and project overhead.

Material quantities, when taken off carefully, can be turned into highly accurate material costs. For this to happen your costs must be current and reliable. The most reliable source for the information is a quote from a reputable vendor for that project. A material supplier should not be chosen on price alone. You must ask if the vendor can deliver the materials at the price quoted. Will they add "special charges" for shipping, delivery, or unloading? Will they be on time? Why are they less expensive? Are there any time limits or escalation clauses on the price? Is this price for the specified materials or is this for an "or equal"? Is this cash up front, C.O.D., or on credit? Is this a stock item that is readily available or will it have to be special ordered? How long will it take? What type of guarantee will they give? Many people would rather pay a higher price to obtain materials from a known, reliable source than risk the unknown. However, there are plenty of bargains out there if you are willing to take some risks.

Other sources of cost information include manufacturers' catalogs (be sure to call and get the contractors' prices or the discount), current cost records from previous jobs, and reputable cost guides.

In order to determine the **labor** costs you need to have three pieces of information: the number of units to be installed, the productivity (how long it takes to install a unit), and the hourly wage. The number of units to be installed are obtained from the material quantity. The productivity of a worker or a crew is very difficult to determine. If you are familiar with a worker or a crew that has been working together for a while, you will have a general "feel" for how much they can install in a day's or week's time. Therefore, the best place to get productivity factors is from your own accurate records from previous projects.

Wage rates are not difficult to find, in most cases. If the project is a union project, then wage rates will be available to you from the union locals. If you are doing an estimate for your own company then you should be able to get the rates from your company's files. You may run into a problem if you are estimating an open shop or non-union project in an unfamiliar location. Often there are organizations in or near larger cities that can assist you in locating this information. Lacking such help, you will have to do some old fashioned "detective work" to get the rates you need.

Equipment costs generally do not include the cost of the operators. In cases where the equipment is listed or leased as an "operated" piece of equipment, the price includes the operator. The costs for rented or leased pieces of equipment can be obtained from local dealers, suppliers, or even manufacturers. These costs can and will fluctuate and need to be updated regularly. Whether you rent or own you will need to add into the overall costs any operating costs: the costs for fuel and lubrication, maintenance (if owned or if in the lease agreement), parts, transportation costs, and mobilization costs. Normally, manufacturers and leasing agents should be able to provide you with these costs.

You can include the equipment costs in two ways. The first is to include the costs as part of the job task; the second is to consider the equipment as part of the overhead of the whole project. The advantage of putting the equipment in with the task is that when reviewing the job records for a future project you will have the equipment costs readily available and thus you will get a clearer picture of the true job cost. Whichever method you choose, be consistent. Make sure that all equipment costs are included and are not duplicated.

Subcontractor quotes should be analyzed in the same fashion as material quotes and the same questions apply. Make sure the subcontractor quote covers **all** the plans and specifications as required. Any exclusions, allowances, alternates, time limits to the bid, and time allowances for work should be duly noted and clearly explained. Keep in mind that you may not look at the estimate until months later. If any subcontractor bids are received by phone, immediately confirm the bid and **all** the details in **writing**. Verbal agreements are subject to a great deal of misunderstanding and misinterpretation.

Project overhead needs to be calculated for this project. Project overhead can be defined as the sum of those items that are necessary for the actual construction of the project, but are not identifiable with any particular work item. The checklist provided in this section specifies some of the items that need to be included. You may not agree with all the items on the list because in some cases you may be able to associate some items with a direct work item. It is not important **where** you place these items as long as you do include them in your estimate.

All the direct costs should be itemized, priced, tabulated, and totaled before the indirect costs are added. Indirect costs are usually added to an estimate at the summary stage and are, in many cases, calculated as a percentage of the direct costs. Included in this grouping are such items as sales tax on materials,

overhead, profit, and contingencies. These indirect costs generally account for the variations among estimates.

Overhead refers to all costs involved in operating your business. It is essential to your company's very existence that these costs be factored into and proportioned over all your projects. The following checklist in this section indicates the expenses that should be covered. The following example indicates how overhead percentages can be computed.

A-Z-Z Contracting, Anytown, USA
Annual Volume = $1,000,000.
Home Office (indirect) Overhead
Calculation — Annual Costs

Salaries

Owner	$40,000
Secretary	15,000
Bookkeeper (part time)	9,000
Accountant (retainer)	4,000
Legal fees (retainer)	4,000
Workers' Comp./medical	7,000
Office	
Rent	8,000
Equipment & utilities	7,000
Supplies	2,000
Advertising	3,000
Auto/Truck	5,000
Association fees	4,000
Entertainment	2,000
Bad debt	5,000
Total	$115,000

Therefore, for each dollar in an estimate, this contractor must add $115,000/ $1,000,000 or 11.5 percent of the total for office overhead.

The following charts, checklists, and data are to be used to help remind you of what items should be included, and their common ranges, in calculating your direct and indirect expenses.

Estimate Summary

After the estimate has been reviewed and all changes have been incorporated, the total for each phase of the work needs to be summarized. Except for estimates with a limited number of items, it is recommended that the costs be transferred from the estimating sheets to an estimate summary sheet. The transferring of the costs should be double-checked, as it is easy to transpose numbers and/or make other simple errors that could cost you thousands of dollars.

The estimate summary sheet should contain the following columns:

    Material
    Labor
    Equipment
    Subcontractor
    Totals

Listing items in their proper columns makes it mathematically simple to arrive at the sum for each category and to apply the appropriate markups to the total dollar values for each column. Each category would normally have different markup percentages added at the end of the estimate for:

    Insurance, taxes, fringe benefits
    Equipment and tools (if equipment is
        being treated as overhead items)
    General overhead, profit,
        contingencies, and fees

The final total on the estimate sheet is the total estimate. If this number is used as the bid figure, you will be committed to it upon acceptance. It is, therefore, very worthwhile to take the time to check that this figure is correct.

A Good Bid

As a bidder, your immediate goal is to submit a bid which is as low as possible but allows you to make a reasonable profit on the job. Therefore you want your bid to be a **good bid**. To define a **good** bid, let's look at the **ideal** bid.

The ideal bid is one that is submitted on work that would be great to get but not a financial loss if you don't, is fair to all parties, allows you to make a good profit, and involves no risk to you. Such a bid is a rarity. In the real world the best we can hope for would be a **good** bid.

A good bid will have the following qualities:

a. be low but not out of line with other bids; in other words, very competitive.

b. involve reasonable to low risk.

c. contain accurate quantities.

d. contain accurate prices from reputable suppliers and subcontractors.

e. provide completeness of coverage.

f. set realistic deadlines.

g. describe realistic plans for execution.

h. include appropriate markups.

i. ensure profitability.

j. utilize properly executed bid forms.

As with any skill, your ability to produce accurate and competitive bids will improve with experience. You can reasonably expect your bidding success rate to improve as your experience grows. Your goal is always to bid low — and show a profit.

Checklist

For an estimate to be reliable, all items must be accounted for. A complete estimate can also eliminate the need to include contingencies. The following checklist can be used to help ensure that all items are properly accounted for.

Direct Overhead Costs

Personnel

- [ ] Superintendent
- [ ] Project Manager (if for that project only)
- [ ] Field Engineer (if for that project only)
- [ ] Cost Engineer (if for that project only)
- [ ] Warehouse personnel (if for that project only)
- [ ] Watchman/guard dogs
- [ ] Tool room keeper (if for that job only)
- [ ] Timekeeper (if for that job only)
- [ ] Foreman (working directly for the contractor)

Temporary Facilities

- [ ] Field office expense
    - ____ Set-up and removal
    - ____ Light
    - ____ Heat
    - ____ Water
    - ____ Telephone
    - ____ Supplies
    - ____ Equipment
        - ____ Fax machine
        - ____ Copy machine
        - ____ Blueprint machine
        - ____ Coffee machine
- [ ] Temporary light and power
- [ ] Temporary heat
- [ ] Temporary water
- [ ] Pay telephones
- [ ] Toilet facilities
- [ ] Enclosures
- [ ] Storage trailers
- [ ] Fencing
- [ ] Barricades and signals
- [ ] Construction road
- [ ] Job sign

Miscellaneous

- [ ] Vehicles
- [ ] Permits
- [ ] Licenses
- [ ] Tools and equipment
- [ ] Photographs
- [ ] Surveying
- [ ] Testing
- [ ] Job signs
- [ ] Pumping
- [ ] Dust control
- [ ] Lifting/hoisting
- [ ] Cleanup (periodical)
- [ ] Final cleanup
- [ ] Damage/repair to adjoining buildings and/or public ways

Indirect Costs

Salaries

- ☐ President
- ☐ Executives
- ☐ Secretaries/Reception
- ☐ Estimators
- ☐ Project Managers
- ☐ Construction Manager
- ☐ Cost Engineers
- ☐ Purchasing Agent
- ☐ Cost/Bookkeeping
- ☐ Engineers
- ☐ Other office personnel
- ☐ Yard personnel
  - ____ Tool Manager
  - ____ Mechanics/Maintenance
  - ____ Drivers
  - ____ Equipment operators

Office

- ☐ Rent/cost of ownership
- ☐ Electricity
- ☐ Gas
- ☐ Water
- ☐ Sewer
- ☐ Telephone
- ☐ Postage
- ☐ Office equipment
- ☐ Furniture/furnishings
- ☐ Office supplies
- ☐ Advertising
- ☐ Literature
- ☐ Club/association dues

Professional Services

- ☐ Legal
- ☐ Accounting
- ☐ Architectural
- ☐ Engineering

Vehicles

- ☐ Cars/trucks
- ☐ Cost of operation
- ☐ Mileage expenses

Insurance

- ☐ Fire
- ☐ Property damage
- ☐ Vehicles
- ☐ Public liability
- ☐ Windstorm
- ☐ Workers' Compensation
- ☐ Unemployment
- ☐ Social Security
- ☐ Flood
- ☐ Theft
- ☐ Elevator

Bonds

- ☐ Bid
- ☐ Payment
- ☐ Performance
- ☐ Surety
- ☐ Lien

Contractor's Overhead & Profit

Below are the **average** installing contractor's percentage mark-ups applied to base labor rates to arrive at typical billing rates.

Column A: Labor rates are based on average open shop wages for 7 major U.S. regions. Base rates including fringe benefits are listed hourly and daily. These figures are the sum of the wage rate and employer-paid fringe benefits such as vacation pay, and employer-paid health costs.

Column B: Workers' Compensation rates are the national average of state rates established for each trade.

Column C: Column C lists average fixed overhead figures for all trades. Included are Federal and State Unemployment costs set at 7.3%; Social Security Taxes (FICA) set at 7.65%; Builder's Risk Insurance costs set at 0.34%; and Public Liability costs set at 1.55%. All the percentages except those for Social Security Taxes vary from state to state as well as from company to company.

Columns D and E: Percentages in Columns D and E are based on the presumption that the installing contractor has annual billing of $500,000 and up. Overhead percentages may increase with smaller annual billing. The overhead percentages for any given contractor may vary greatly and depend on a number of factors, such as the contractor's annual volume, engineering and logistical support costs, and staff requirements. The figures for overhead and profit will also vary depending on the type of job, the job location, and the prevailing economic conditions. All factors should be examined very carefully for each job.

Column F: Column F lists the total of Columns B, C, D, and E.

Column G: Column G is Column A (hourly base labor rate) multiplied by the percentage in Column F (O&P percentage).

Column H: Column H is the total of Column A (hourly base labor rate) plus Column G (Total O&P).

Column I: Column I is Column H multiplied by eight hours.

		A		B	C	D	E	F	G	H	I
		Base Rate Incl. Fringes		Workers' Comp. Ins.	Average Fixed Over-head	Over-head	Profit	Total Overhead & Profit		Rate with O & P	
Abbr.	Trade	Hourly	Daily					%	Amount	Hourly	Daily
Skwk	Skilled Workers Average (35 trades)	$17.05	$136.40	20.2%	16.8%	27.0%	10%	74.0%	$12.60	$29.65	$237.20
	Helpers Average (5 trades)	12.65	101.20	21.4		25.0		73.2	9.25	21.90	175.20
	Foreman Average, Inside ($.50 over trade)	17.55	140.40	20.2		27.0		74.0	13.00	30.55	244.40
	Foreman Average, Outside ($2.00 over trade)	19.05	152.40	20.2		27.0		74.0	14.10	33.15	265.20
Clab	Common Building Laborers	12.30	98.40	21.9		25.0		73.7	9.05	21.35	170.80
Asbe	Asbestos Workers	18.00	144.00	19.7		30.0		76.5	13.75	31.75	254.00
Boil	Boilermakers	18.95	151.60	17.7		30.0		74.5	14.10	33.05	264.40
Bric	Bricklayers	17.35	138.80	19.4		25.0		71.2	12.35	29.70	237.60
Brhe	Bricklayer Helpers	13.40	107.20	19.4		25.0		71.2	9.55	22.95	183.60
Carp	Carpenters	16.90	135.20	21.9		25.0		73.7	12.45	29.35	234.80
Cefi	Cement Finishers	16.30	130.40	12.8		25.0		64.6	10.55	26.85	214.80
Elec	Electricians	18.75	150.00	8.0		30.0		64.8	12.15	30.90	247.20
Elev	Elevator Constructors	19.25	154.00	9.6		30.0		66.4	12.80	32.05	256.40
Eqhv	Equipment Operators, Crane or Shovel	17.65	141.20	12.9		28.0		67.7	11.95	29.60	236.80
Eqmd	Equipment Operators, Medium Equipment	16.95	135.60	12.9		28.0		67.7	11.50	28.45	227.60
Eqlt	Equipment Operators, Light Equipment	16.30	130.40	12.9		28.0		67.7	11.05	27.35	218.80
Eqol	Equipment Operators, Oilers	14.45	115.60	12.9		28.0		67.7	9.80	24.25	194.00
Eqmm	Equipment Operators, Master Mechanics	18.20	145.60	12.9		28.0		67.7	12.30	30.50	244.00
Glaz	Glaziers	16.95	135.60	16.0		25.0		67.8	11.50	28.45	227.60
Lath	Lathers	16.70	133.60	13.5		25.0		65.3	10.90	27.60	220.80
Marb	Marble Setters	17.20	137.60	19.4		25.0		71.2	12.25	29.45	235.60
Mill	Millwrights	17.80	142.40	13.2		25.0		65.0	11.55	29.35	234.80
Mstz	Mosaic & Terrazzo Workers	16.90	135.20	11.0		25.0		62.8	10.60	27.50	220.00
Pord	Painters, Ordinary	15.60	124.80	16.8		25.0		68.6	10.70	26.30	210.40
Psst	Painters, Structural Steel	16.30	130.40	62.5		25.0		114.3	18.65	34.95	279.60
Pape	Paper Hangers	15.85	126.80	16.8		25.0		68.6	10.85	26.70	213.60
Pile	Pile Drivers	17.00	136.00	33.6		30.0		90.4	15.35	32.35	258.80
Plas	Plasterers	16.20	129.60	17.4		25.0		69.2	11.20	27.40	219.20
Plah	Plasterer Helpers	13.50	108.00	17.4		25.0		69.2	9.35	22.85	182.80
Plum	Plumbers	18.95	151.60	10.2		30.0		67.0	12.70	31.65	253.20
Rodm	Rodmen (Reinforcing)	17.75	142.00	36.3		28.0		91.1	16.15	33.90	271.20
Rofc	Roofers, Composition	14.90	119.20	37.4		25.0		89.2	13.30	28.20	225.60
Rots	Roofers, Tile & Slate	14.90	119.20	37.4		25.0		89.2	13.30	28.20	225.60
Rohe	Roofers, Helpers (Composition)	10.55	84.40	37.4		25.0		89.2	9.40	19.95	159.60
Shee	Sheet Metal Workers	18.25	146.00	13.8		30.0		70.6	12.90	31.15	249.20
Spri	Sprinkler Installers	19.70	157.60	10.4		30.0		67.2	13.25	32.95	263.60
Stpi	Steamfitters or Pipefitters	19.10	152.80	10.2		30.0		67.0	12.80	31.90	255.20
Ston	Stone Masons	16.85	134.80	19.4		25.0		71.2	12.00	28.85	230.80
Sswk	Structural Steel Workers	17.80	142.40	46.4		28.0		101.2	18.00	35.80	286.40
Tilf	Tile Layers	16.80	134.40	11.0		25.0		62.8	10.55	27.35	218.80
Tilh	Tile Layers Helpers	13.60	108.80	11.0		25.0		62.8	8.55	22.15	177.20
Trlt	Truck Drivers, Light	13.85	110.80	17.0		25.0		68.8	9.55	23.40	187.20
Trhv	Truck Drivers, Heavy	14.10	112.80	17.0		25.0		68.8	9.70	23.80	190.40
Sswl	Welders, Structural Steel	17.80	142.40	46.4		28.0		101.2	18.00	35.80	286.40
Wrck	*Wrecking	12.65	101.20	44.8		25.0		96.6	12.20	24.85	198.80

*Not included in Averages.

General Contractor's Overhead

The table below shows a contractor's overhead as a percentage of direct cost in two ways. The figures on the right are for the overhead, markup based on both material and labor. The figures on the left are based on the entire overhead applied only to the labor. This figure would be used if the owner supplied the materials or if a contract is for labor only.

Items of General Contractor's Indirect Costs	% of Direct Costs	
	As a Markup of Labor Only	As a Markup of Both Material and Labor
Field Supervision	6.0%	3.2%
Main Office Expense (see details below)	14.7	7.7
Tools and Minor Equipment	1.0	0.5
Workers' Compensation & Employers' Liability.	20.2	10.6
Field Office, Sheds, Photos, Etc.	1.5	0.8
Performance and Payment Bond, 0.5% to 0.9%.	0.7	0.4
Unemployment Tax See R010-100 (Combined Federal and State)	7.3	3.8
Social Security and Medicare (7.65% of first $61,200)	7.7	4.0
Sales Tax — add if applicable 38/80 x % as markup of total direct costs including both material and labor.		
Sub Total	59.1%	31.0%
*Builder's Risk Insurance ranges from .141% to .586%.	0.3	0.3
*Public Liability Insurance	1.5	1.5
Grand Total	60.9%	32.8%

*Paid by Owner or Contractor

General Contractor's Main Office Expense for Smaller Projects

This table provides average main office expenses (as a percentage of annual volume) for contractors specializing in smaller projects, such as repair and/or remodeling of existing structures.

Annual Volume	% of Annual Volume	
	Minimum	Maximum
To $50,000	20%	30%
To $100,000	17%	22%
To $250,000	16%	19%
To $500,000	14%	16%
To $1,000,000	8%	10%

Overhead and Profit as a Percentage of Project Costs

Overhead is defined as costs that are associated with a construction project, but not directly with the actual construction. This table shows percentages that can be used as a rule of thumb for estimating overhead costs.

Overhead as a Percentage of Direct Costs		Overhead and Profit Allowance — Add to Items That Do Not Include Subcontractor's O&P — Average	Allowance to Add to Items That Do Include Subcontractor's O&P		Typical by Size of Project	
Minimum	5%	25%	Minimum	5%	under $100,000	30%
Average	12%		Average	10%	$500,000	25%
Maximum	22%		Maximum	15%	$2,000,000	20%
					over $10,000,000	15%

Insurance Rates

Type	Minimum	Maximum
Builder's Risk	.22%	.59%
All-risk Type	.25%	.62%
Contractor's Equipment Floater	.50%	1.50%
Public Liability, Average	—	1.55%

This table represents approximate values relative to total project cost for the most common types of basic insurance coverages.

Builder's Risk Insurance Rates

Builder's Risk Insurance is insurance on a building during construction. Premiums are paid by the owner or the contractor. Blasting, collapse and underground insurance would raise total insurance costs above those listed. Floater policy for materials delivered to the job runs $.75 to $1.25 per $100 value. Contractor equipment insurance runs $.50 to $1.50 per $100 value. Insurance for miscellaneous tools to $1,000 value runs from $5.50 to $7.50 per $100 value.

Tabulated below are New England Builder's Risk insurance rates in dollars per $100 value for $1,000 deductible. For $25,000 deductible, rates can be reduced 13% to 34%. On contracts over $1,000,000, rates may be lower than those tabulated. Policies are written annually for the total completed value in place. For "all risk" insurance (excluding flood, earthquake and certain other perils) add $.025 to total rates below.

Coverage	Frame Construction (Class 1)			Brick Construction (Class 4)			Fire Resistive (Class 6)		
	Range		Average	Range		Average	Range		Average
Fire Insurance	$.300 to $.420		$.394	$.132 to $.189		$.174	$.052 to $.080		$.070
Extended Coverage	.115 to .150		.144	.080 to .105		.101	.081 to .105		.100
Vandalism	.012 to .016		.015	.008 to .011		.011	.008 to .011		.010
Total Annual Rate	$.427 to $.586		$.553	$.220 to .305		$.286	$.141 to $.196		$.180

Permit Rates

Project Permits	Minimum	Maximum
Rule of thumb, most cities	.50%	2%

Permit costs vary greatly depending on many factors, such as type of project, proposed occupancy, location, local codes, need for variances or change of zoning, etc. This table provides a "rule of thumb" to use when local conditions cannot be determined.

Small Tools Allowance

Small Tools Allowance	Minimum	Maximum
As % of contractor's work	.50%	2%

A variety of small tools must be purchased in the course of almost every project, whether to complete small tasks or to replace tools that have "mysteriously disappeared." This table provides a "rule of thumb" that can be used to assign a value to this often overlooked item.

Project Desirability

In the estimating and bidding of construction projects, one of the first questions that you must ask yourself is "Do I really want to bid this job?" The question is simple enough; it's the answer that is sometimes the problem. Judging whether you should bid on a project or pass on it depends on its desirability as viewed by each bidder. Some projects are clear cut in this regard. Those that fall somewhere in the middle are hard to call.

Bid dates (deadlines), dollar size, site location, type of construction, and other pertinent conditions affect the immediate choice. They will continue to influence the deeper analysis that must follow.

Analyzing the project serves three purposes: (1) to reach a decision on whether or not to bid, (2) to determine the degree of effort and competitiveness to apply if you decide to bid, and (3) to establish the amount of the final markup for profit.

In reality, a bidder performs this analysis, referred to hereafter as a project's rating, or desirability rating, somewhat intuitively and with little conscious deliberation. A few typical conditions are:

1. Size (cost of the project in dollars)

2. Location of project (distance of the project from the home office)

3. Sponsor relation

4. Type of construction

5. Probable competitiveness

6. Labor market

7. Subcontractor market

8. Quality of drawings and specifications

9. Quality of supervision

10. Special risks

11. Completion time and penalty

12. Estimating and bidding time

13. Need for work

14. Other special advantages or disadvantages

If all these conditions were positive, the project would be highly desirable, the bidding competitive, and the markup minimal. If most of these conditions were negative, the project would be less desirable, the bidding conservative, and the markup maximum.

The markup is a variable that helps to compensate for the strengths or weaknesses of a project, as summarized in its rating. For instance, the example below shows two projects of equal size with different ratings. The increased markup tends to make Project A more desirable. And yet, if the theory is correct, it would not seriously decrease competitiveness, since other bidders should also make similar compensations for the obvious differences.

	Basic Cost	Rating	Markup	Add'l. Cost
Project A	$42,000	78	*10.2%	$4,284
Project B	$42,000	88	*9.0%	$3,780

*Assumed Theoretical Percentages

This example graphically portrays the principle.

If both projects were equal in ratings, the markups and bids would be equal. The rating works both ways, to raise or to lower a bid relative to some arbitrary average. It is thus a realistic aspect of good bidding practice.

Let us now examine each of the previously named conditions. We cannot expect them to be of equal weight, so there should be different rules for each. Let us reserve all numbers on the 0-100 scale below 30 and above 90 for very exceptional conditions, in order to give them additional impact in the final averaging.

1. **Size of project.** The ideal size is that which will fill up the company's unused bonding capacity. If, for instance, a company has a nominal bonding capacity of $200,000 and $120,000 of uncompleted work on hand, then an $80,000 project would be an ideal size for bidding.

   Let us rate projects for size between the two following extremes:

Ideal size	90
Smaller than ideal	50

2. **Location of project.** For the distance of the project from the home office, suggested values are:

Less than 5 miles	90
More than 30 miles	60

3. **Owner relations.** Experienced bidders know that the efficiency and cooperativeness of the owner's agents contribute to the desirability rating of a project and can influence the decision of the bidder on whether to bid or not. Suggested values are:

Known good relations	90
Unknown	70
Known poor relations	50

4. **Type of construction.** As a rule, a contractor is not equally experienced and capable in all types of construction. The type of construction in which the company excels is the most desirable to a bidder and determines the rating between the following extremes:

Experienced in this type	90
No experience in this type	60

5. **Probable.** The bidder looks with less favor on those projects that attract a large number of bidders and excessive competition. This aspect of a project can carry a lot of weight with values suggested as follows:

Light competition	90
Strong competition	60

6. **Labor market conditions.** Projects located in areas where there are an insufficient number of skilled workers are of low desirability to bidders. Suggested values are:

Good quality and plentiful	90
Poor quality and few	60

7. **Subcontract market conditions.** The quantity and quality of available local subcontractors parallels that of workers, and the suggested values are:

Good quality and plentiful	90
Poor quality and few	60

8. **Quality of drawings and specifications.** Quality is not a heavy factor, but it does affect the desirability of a project. Suggested values are:

Good quality	90
Poor quality	70

9. **Quality of supervision.** When the identity of the superintendent is known and of proven excellence, the project is more desirable to the bidder. An unknown superintendent, or one of mediocre ability, is a negative influence in the rating. The following values are suggested:

Known top quality	90
Unknown	80
Known mediocre quality	70

10. **Special risks.** Conditions such as solid rock, subsurface water, high altitude, extreme heat or cold, possibility of collapse, cave-in, etc., all affect the desirability rating. Depending on the degree of risk, values may be scaled between:

No known risk	90
Extreme risk	30

11. **Completion time and penalty.** When the time provided in the bid documents to complete the construction is extremely short and the penalty for delay is very high, the project loses desirability to the bidder. Values are:

Time sufficient and penalty low	90
Time sufficient and penalty high	60

12. **Estimating and bidding time.** If the bidder is not given sufficient time to put the bid together, a low level of desirability is assigned to the project.

This condition might also exist because of the coincidence of two projects bidding on or near the same date. Values are:

Sufficient time to bid	90
Insufficient time to bid	50

13. **Need for work.** This condition could carry more weight than any of the others. If extreme, the bidder's need for work might reverse an otherwise unfavorable rating. Suggested values are:

Extreme need for work	100
Normal need	85
Very little need	50

14. **Other special advantages or disadvantages.** A bidder might have special incentives, positive or negative, such as the exclusive possession of a stockpile of materials. Or the bidder might have knowledge of a competitor's possession of such an advantage. The weight for this condition can vary considerably, but if there are no special advantages or disadvantges, an average may be used, as follows:

Strong advantages	100
Average	80
Strong disadvantages	25

In a similar way, but with fewer facts, the bidder roughly analyzes the competition's probable degree of interest in the project. The result may cause a further final adjustment in the bidder's own evaluation.

The following figure is an example of a hypothetical project rated for desirability.

1. Size of project in dollars	90
2. Location of project (distance from home office)	90
3. Owner, architect, inspector (sponsor) relations	80
4. Type of construction (experienced or not)	90
5. Competition	80
6. Labor market	75
7. Subcontractor market	80
8. Quality of drawings and specifications	85
9. Quality of supervision	90
10. Special risks	25
11. Completion time and liquidated damages	85
12. Time to estimate and bid	80
13. Need of a job (contract)	100
14. Other	85
	1,135

$$\frac{1,135}{14} = 81.07 \text{ average}$$

Two extremes are shown: special risk and need for work. These two exceptional conditions tend to balance one another and make the project of average desirability, according to the following scale:

Very desirable	85 to 90
Average	80 to 85
Undesirable	Below 80

Remember, a project may be made more desirable by an increase in the markup and by contingency allowances.

Basics of Project Management

Anyone who has spent an hour on a job site, from the seasoned old pro to the greenest young laborer, knows that any construction project is a monument to Murphy's Law, ("Whatever **can** go wrong, **will** go wrong."), and its corollary, ("It will go wrong at the worst possible time."). A construction project can be viewed as an ongoing exercise in creative problem-solving on the part of everyone involved, with the buck ultimately stopping on the desk of the project manager. It can be truly said that success in the construction business is in direct correlation to your ability to master the skills of project management.

Construction project management can be defined as the planning, staffing, directing, and controlling of a company's resources in order to achieve the goals set for a construction project. What follows is a brief outline of the basic tasks involved in successful project management.

1. **Estimate** as accurately as possible the cost of the project. This should include a detailed breakdown of all phases ("systems") of the project in terms of time ("labor-hours") and materials ("components"). No management skill has more of a direct impact on your profitability than your ability to produce an accurate estimate.

2. **Establish specific goals** in terms of how much time and money you wish to spend. These do not necessarily have to correspond exactly to your estimate. For example, your estimate might call for framing to be completed in six weeks, but you set your goal for five weeks. Your overall goal is always not to exceed your estimate. Coming in under your estimate is money in your pocket.

3. **Plan ahead** on how you are going to achieve your goals. This may involve anything from deciding how stock is to be arranged on the job site to coordinating the purchase of materials to take advantage of price breaks offered by different suppliers. Planning takes time and effort, but it is as much a part of a successful builder's skills as the ability to drive a nail. Anticipating problems and solving them through careful and imaginative planning is the most effective way of ensuring that the physical job of building will go smoothly and result in good work at minimal cost. You cannot completely escape the clutches of Murphy's Law, but good planning practices will generally bail you out.

4. **Communicate** your goals and plans to those who will execute or be affected by them. Be certain your crew understands what you expect them to produce and how and when you expect them to produce it. Give your directions clearly and encourage them to ask questions whenever they are unsure of what their job is or how it is to be done. Keep your client up to date on the progress of the job and solicit his or her comments and suggestions, particularly regarding design and finish treatment. This can often be an annoying task and a trial of your patience, but keeping your client involved in the process and satisfied

with the results is a very important factor in your formula for success. The client, after all, writes your paycheck and represents the best, and cheapest, advertisements for your business.

5. **Establish** a chain of command. Be sure everyone involved in the project understands who is in charge of, or answerable to, whom. From architects and designers, through foremen and subcontractors, down to the laborer who sweeps up the sawdust — all must be aware of how decisions are made and who is called upon to make them. This is often a self-evident matter within the various crews, but it can get extremely complicated when a problem or design change comes up which may involve one architect, two engineers, and three tradesmen, all of whom seem to be at cross-purposes. You must then bring to bear all your skills as a communicator, mediator, and problem-solver to come to a decision that satisfies everyone.

6. **Coordinate** the use of resources to meet your goals. Simply put, this means putting your plans into action. You have spent time deciding on the scheduling of operations, the building sequence, and the most efficient and economical use of materials, labor, tools, equipment, and outside services. Now it is up to you to see that everything gets done as planned. The success of your project very much depends on taking those ideas you put on paper and putting them into practice on the job. After all, there is no sense in writing a stock list if you don't read it when you get to the lumberyard.

7. **Monitor and document** all activities and events that affect, directly or indirectly, the work that you and/or your company control. Maintain a project journal in a notebook. Set aside a little time each day to record a brief account of the job's progress: goals and deadlines met or missed;

problems confronted: their causes and solutions; personnel difficulties; business conversations; promises or commitments received or given; in short, anything of significance to the orderly completion of the project. You will thus have, for immediate reference, an up-to-date assessment of the state of the project, as well as the information necessary to make intelligent decisions and adjustments to keep things running smoothly. Journals of past projects can shed light on current problems and will often contain insights, answers, solutions, and information you forgot you knew. As part of the same journal, or perhaps in a separate notebook, maintain an updated list of "extras," that is, labor and materials not included in the original estimate. Write down a description of each extra, noting costs, and have it initialed by the client, designer, subcontractor, or any other person in charge of carrying out the work. This ensures that all interested parties are aware and informed of all extras and of their specific obligations with regard to them. Confusion, misunderstanding, and hard feelings are thus forestalled, and a professional tone is maintained on the job that will be to everyone's benefit and help to produce quality results.

Overall project management can be more fully illustrated in the following figure — the Project Control Cycle. Notice that the feedback loop is part of the overall project management responsibility.

The steps involved in the Project Control Cycle are:

STEP 1 — Set Initial Goals. This involves creating an estimate of cost and time needed for the project. In many cases these are already established for you by way of a bid and/or a contract.

STEP 2 — Establish the Job Plan. Build a schedule and create a budget of how much you want to spend — not how much you can spend.

STEP 3 — Monitor the Progress. Collect data on how much money has been spent, how much time has been used, and the overall state of the job.

STEP 4 — Process the Information. Compare the data collected in Step 3 against the job plan.

STEP 5 — Compare and Analyze. Look for deviations from the job plan. Determine what the causes are.

STEP 6 — Take Corrective Action. Decide what needs to be done and do it, or have it done. (NOTE: If you do not take corrective action when it is warranted then the whole concept of project management falls apart. If you are not prepared to take corrective action, no management system will work.)

STEP 7 — Collect Historical Data. The information gathered on any project serves many purposes. It will: 1) be extremely useful when estimating your next job; 2) help prevent you from falling into a bad situation again; 3) assist you in getting out of a similar situation if it should occur again; and 4) document all events clearly as they occur, in case legal action is taken by or against you.

We cannot guarantee that all of your projects will be successful by using project management procedures, but, if used properly, such procedures will go a long way toward making your projects run smoother and your business more profitable.

Scheduling Concerns

There are many factors that must be taken into consideration when preparing a construction schedule. They include the time allowed (in a contract, for instance), job location, the climate of the region, the time of the year, deliveries and availabilities of materials, the crew make-up, size and complexity of the job, administrative requirements, sample and document submittals for approval, and the coordination with other trades.

Scheduling is determining the sequence of events that must occur to complete your project. To create a schedule, you must examine each task of your project and ask these three questions:

What task must be completed before this activity can start?
What task can be worked at the same time as this task?
What task cannot start until this activity is complete?

When the order of work is determined by the above process, you have a basic schedule completed. You now need to figure out how long each task will take to complete. Work with your superintendents, foremen and/or the workers. They have the working knowledge of how long it takes to complete a task. If you are doing work that you and/or your crews are not familiar with, then consult the Means labor-hour columns to give you a good idea of how long it will take.

If the time allowed for you to complete the work is already spelled out for you, you at least have a framework of time to work within and can form your crews to fit accordingly.

Allow for weather delays in regions and/or times of year they are warranted. In New England, for outside work as a general rule, add two days per week for winter work, add one day per week for the spring and the fall seasons, and one day for every two weeks of work in the summertime.

You also need to take into consideration the availability of workers in the trades; whether they live near the project; whether they'll stay on or jump to another job, or whether they have experience in this type of work.

A last item concerning scheduling: Do not let the schedule absolutely dictate how the job is to be run. If a long lead item suddenly shows up on a site a week early, don't determine that it has to sit in a storage trailer until the schedule says it's okay to install it. These are the pleasant type of surprises we rarely get! Rework your schedule to determine if you can save any time by installing it **now**. Again, run through various "what if...?" types of situations. In other words, be firm — maintain the schedule at all costs; but be flexible enough and open to the inevitable changes that will occur on your job site. If your information is current and logic is sound, you should be able to take advantage of any situation that occurs and turn it into a gain for yourself and your project.

Construction Forms

Forms are an important management tool. Consistent use of a specific set of forms helps to organize and standardize the collection and reporting of information. As a result, those who are involved in data gathering and analysis become so familiar with the use of the forms that they can more easily focus on the information without being distracted by the form itself. Estimate forms are designed to record and summarize the quantity calculations and systems pricing on a single sheet. There are columns to list descriptive information as well as material and installation costs on a per unit basis. Additional columns are provided to calculate project specific costs after determination of total square footage or total number of units appropriate to the system being estimated. Totals are then determined by adding the extended figures across and down the form.

The forms in this section are ready for direct use. After the appropriate form has been selected, the form can be reproduced on an office copier as needed. Purchase of this book grants the owner permission to reproduce unlimited quantities of these forms for his/her own use but not for resale.

Sample Form

Floor Framing System

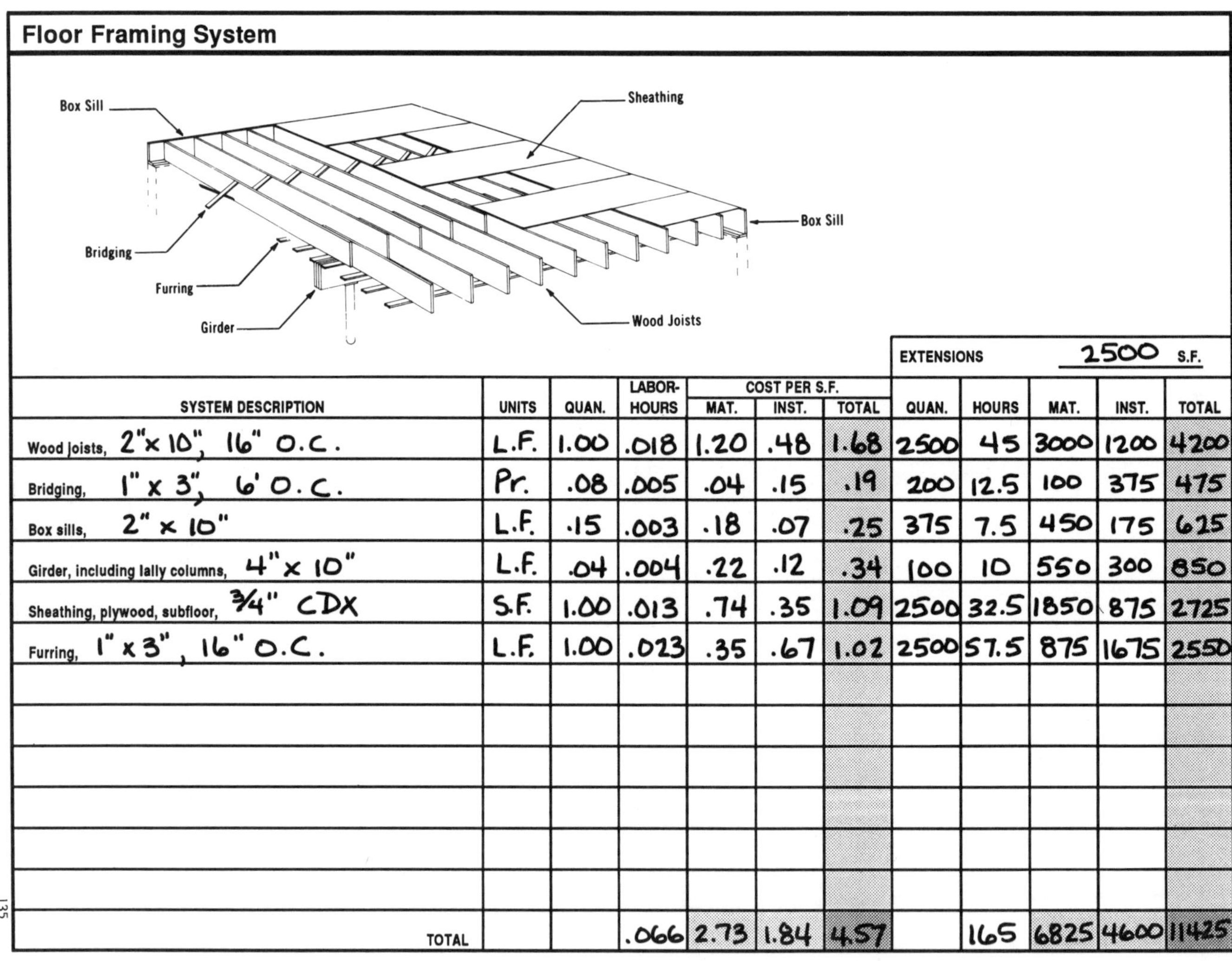

EXTENSIONS   2500 S.F.

SYSTEM DESCRIPTION	UNITS	QUAN.	LABOR-HOURS	COST PER S.F.			QUAN.	HOURS	MAT.	INST.	TOTAL
				MAT.	INST.	TOTAL					
Wood joists, 2" x 10", 16" O.C.	L.F.	1.00	.018	1.20	.48	1.68	2500	45	3000	1200	4200
Bridging, 1" x 3", 6' O.C.	Pr.	.08	.005	.04	.15	.19	200	12.5	100	375	475
Box sills, 2" x 10"	L.F.	.15	.003	.18	.07	.25	375	7.5	450	175	625
Girder, including lally columns, 4" x 10"	L.F.	.04	.004	.22	.12	.34	100	10	550	300	850
Sheathing, plywood, subfloor, 3/4" CDX	S.F.	1.00	.013	.74	.35	1.09	2500	32.5	1850	875	2725
Furring, 1" x 3", 16" O.C.	L.F.	1.00	.023	.35	.67	1.02	2500	57.5	875	1675	2550
TOTAL			.066	2.73	1.84	4.57		165	6825	4600	11425

Floor Framing System

| SYSTEM DESCRIPTION | UNITS | QUAN. | LABOR-HOURS | COST PER S.F. | | | EXTENSIONS QUAN. | HOURS | MAT. | INST. | TOTAL |
				MAT.	INST.	TOTAL					S.F.
Wood Joists,											
Bridging,											
Box sills,											
Girder, including lally columns,											
Sheathing, plywood, subfloor,											
Furring,											
TOTAL											

Floor Framing System

SYSTEM DESCRIPTION	UNITS	QUAN.	LABOR-HOURS	COST PER S.F.			EXTENSIONS				S.F.
				MAT.	INST.	TOTAL	QUAN.	HOURS	MAT.	INST.	TOTAL
Composite Wood Joists											
Temp. strut line											
CWJ rim joist											
Girder, including lally columns,											
Sheathing, plywood, subfloor,											
TOTAL											

Floor Framing System

SYSTEM DESCRIPTION	UNITS	QUAN.	LABOR-HOURS	COST PER S.F.			EXTENSIONS				
				MAT.	INST.	TOTAL	QUAN.	HOURS	MAT.	INST.	TOTAL
Open Web Wood joists											
Continuous ribbing											
Girder, including lally columns,											
Sheathing, plywood, subfloor,											
Furring,											
TOTAL											

Exterior Wall Framing System

SYSTEM DESCRIPTION	UNITS	QUAN.	LABOR-HOURS	COST PER S.F.			EXTENSIONS				S.F.
				MAT.	INST.	TOTAL	QUAN.	HOURS	MAT.	INST.	TOTAL
Wood studs,											
Plates, double top, single bottom,											
Corner bracing,											
Sheathing,											
TOTAL											

Gable End Roof Framing System

SYSTEM DESCRIPTION	UNITS	QUAN.	LABOR-HOURS	COST PER S.F.			EXTENSIONS				S.F.
				MAT.	INST.	TOTAL	QUAN.	HOURS	MAT.	INST.	TOTAL
Wood rafters,											
Ceiling joists,											
Ridge board,											
Fascia board,											
Rafter tie,											
Soffit nailer,											
Sheathing,											
Furring strips,											
TOTAL											

Truss Roof Framing System

| SYSTEM DESCRIPTION | UNITS | QUAN. | LABOR-HOURS | COST PER S.F. | | | EXTENSIONS | | | | S.F. |
				MAT.	INST.	TOTAL	QUAN.	HOURS	MAT.	INST.	TOTAL
Truss, loading,											
Fascia board,											
Sheathing,											
Furring,											
TOTAL											

Hip Roof Framing System

| SYSTEM DESCRIPTION | UNITS | QUAN. | LABOR-HOURS | COST PER S.F. | | | EXTENSIONS | | | | S.F. |
				MAT.	INST.	TOTAL	QUAN.	HOURS	MAT.	INST.	TOTAL
Hip rafters,											
Jack rafters,											
Ceiling joists,											
Fascia board,											
Soffit nailer,											
Sheathing,											
Furring strips,											
TOTAL											

Gambrel Roof Framing System

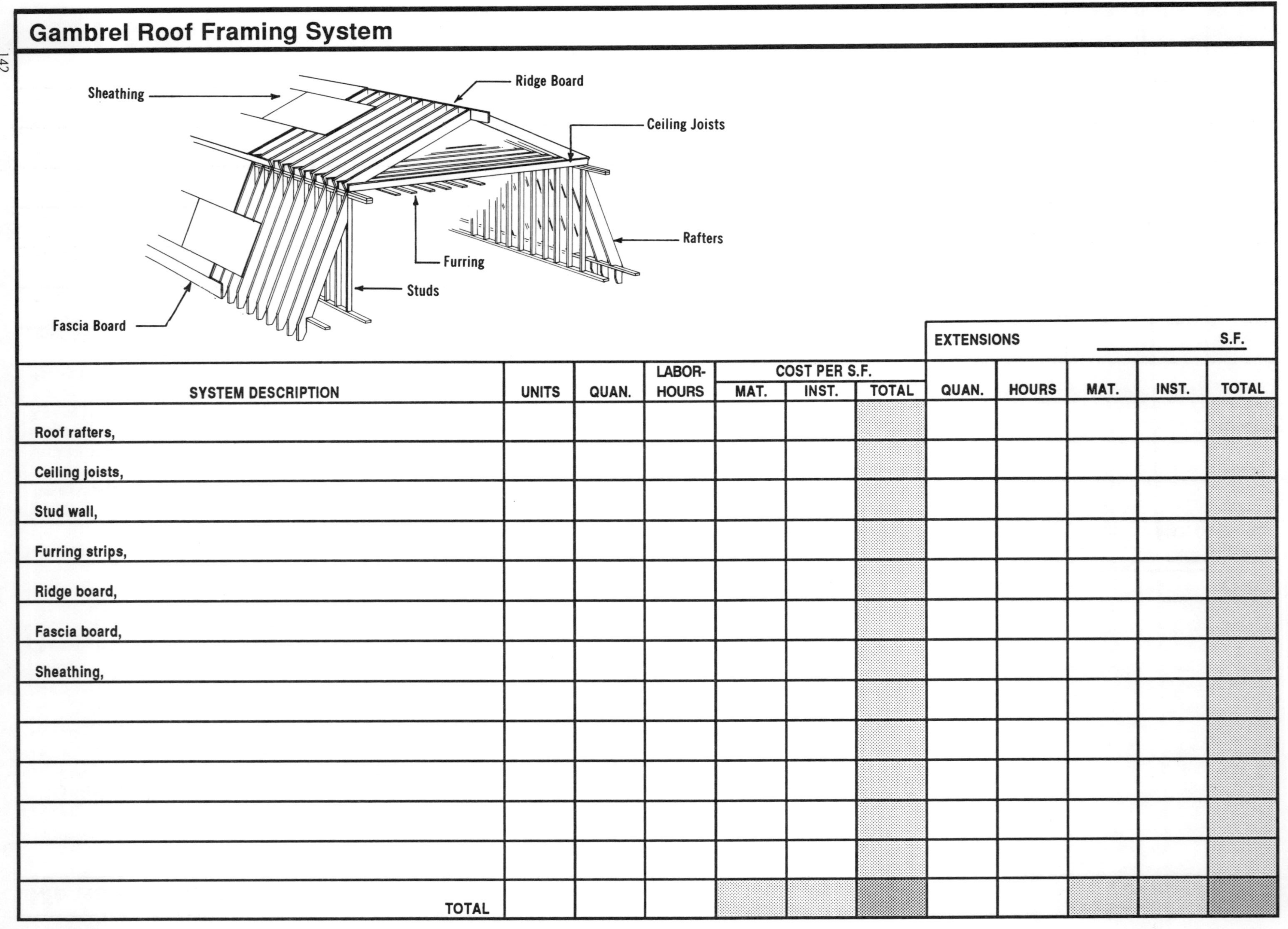

SYSTEM DESCRIPTION	UNITS	QUAN.	LABOR-HOURS	COST PER S.F.			EXTENSIONS				S.F.
				MAT.	INST.	TOTAL	QUAN.	HOURS	MAT.	INST.	TOTAL
Roof rafters,											
Ceiling joists,											
Stud wall,											
Furring strips,											
Ridge board,											
Fascia board,											
Sheathing,											
TOTAL											

Mansard Roof Framing System

SYSTEM DESCRIPTION	UNITS	QUAN.	LABOR-HOURS	COST PER S.F.			EXTENSIONS QUAN.	HOURS	MAT.	INST.	TOTAL
				MAT.	INST.	TOTAL					
Roof rafters,											
Rafter plates,											
Ceiling joists,											
Hip rafter,											
Jack rafter,											
Ridge board,											
Sheathing,											
Furring strips,											
TOTAL											

Shed/Flat Roof Framing System

SYSTEM DESCRIPTION	UNITS	QUAN.	LABOR-HOURS	COST PER S.F.			EXTENSIONS QUAN.	HOURS	MAT.	INST.	TOTAL
				MAT.	INST.	TOTAL					
Roof rafters,											
Fascia board,											
Bridging,											
Sheathing,											
TOTAL											

Gable Dormer Framing System

SYSTEM DESCRIPTION	UNITS	QUAN.	LABOR-HOURS	COST PER S.F.			EXTENSIONS QUAN.	HOURS	MAT.	INST.	TOTAL
				MAT.	INST.	TOTAL					S.F.
Dormer rafter,											
Ridge board,											
Trimmer rafters,											
Wall, studs and plates,											
Fascia board,											
Valley rafter,											
Cripple rafter,											
Headers,											
Ceiling joists,											
Sheathing,											
TOTAL											

Shed Dormer Framing System

SYSTEM DESCRIPTION	UNITS	QUAN.	LABOR-HOURS	COST PER S.F. MAT.	COST PER S.F. INST.	COST PER S.F. TOTAL	EXTENSIONS QUAN.	HOURS	MAT.	INST.	TOTAL
Dormer rafter,											
Trimmer rafter,											
Wall, studs and plates,											
Fascia board,											
Ceiling joists,											
Sheathing,											
TOTAL											

Partition Framing System

| SYSTEM DESCRIPTION | UNITS | QUAN. | LABOR-HOURS | COST PER S.F. | | | EXTENSIONS | | | | S.F. |
				MAT.	INST.	TOTAL	QUAN.	HOURS	MAT.	INST.	TOTAL
Wood studs,											
Plates, double top, single bottom,											
Cross bracing,											
TOTAL											

Wood Siding System

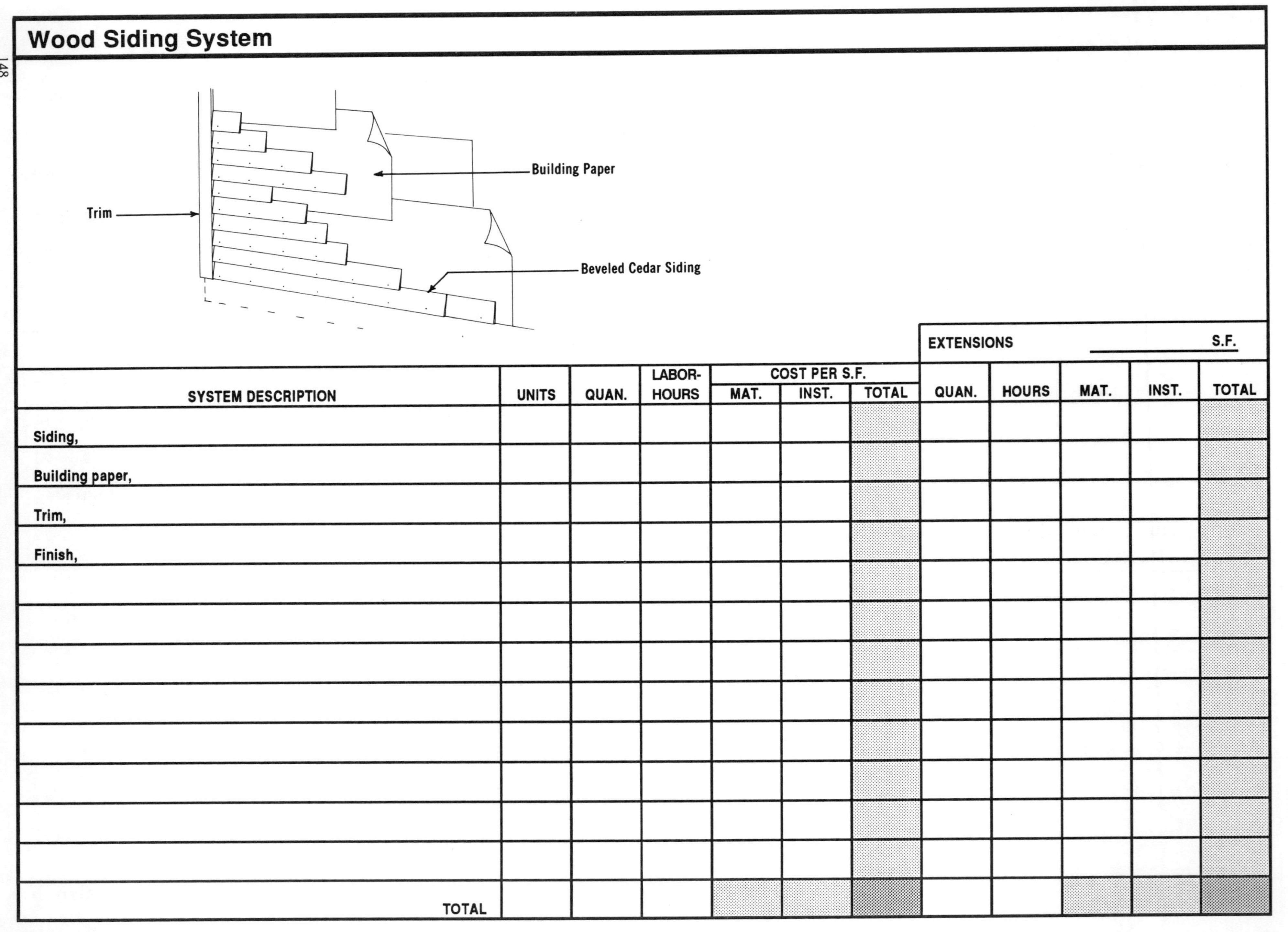

SYSTEM DESCRIPTION	UNITS	QUAN.	LABOR-HOURS	COST PER S.F.			EXTENSIONS				S.F.
				MAT.	INST.	TOTAL	QUAN.	HOURS	MAT.	INST.	TOTAL
Siding,											
Building paper,											
Trim,											
Finish,											
TOTAL											

Shingle Siding System

Trim

Building Paper

White Cedar Shingles

SYSTEM DESCRIPTION	UNITS	QUAN.	LABOR-HOURS	COST PER S.F.			EXTENSIONS				S.F.
				MAT.	INST.	TOTAL	QUAN.	HOURS	MAT.	INST.	TOTAL
Shingles,											
Building paper,											
Trim,											
Finish,											
TOTAL											

Metal and Plastic Siding System

SYSTEM DESCRIPTION	UNITS	QUAN.	LABOR-HOURS	COST PER S.F.			EXTENSIONS				S.F.
				MAT.	INST.	TOTAL	QUAN.	HOURS	MAT.	INST.	TOTAL
Siding,											
Backer, insulation board,											
Building paper,											
Trim,											
TOTAL											

Double Hung Window System

| SYSTEM DESCRIPTION | UNITS | QUAN. | LABOR-HOURS | COST EACH | | | EXTENSIONS | | | | EACH |
				MAT.	INST.	TOTAL	QUAN.	HOURS	MAT.	INST.	TOTAL
Window,											
Trim,											
Finish,											
Caulking,											
Grille,											
Drip cap,											
TOTAL											

Casement Window System

SYSTEM DESCRIPTION	UNITS	QUAN.	LABOR-HOURS	COST EACH			EXTENSIONS				EACH
				MAT.	INST.	TOTAL	QUAN.	HOURS	MAT.	INST.	TOTAL
Window,											
Trim,											
Finish,											
Caulking,											
Grille,											
Drip cap,											
TOTAL											

Awning Window System

SYSTEM DESCRIPTION	UNITS	QUAN.	LABOR-HOURS	COST EACH			EXTENSIONS			EACH	
				MAT.	INST.	TOTAL	QUAN.	HOURS	MAT.	INST.	TOTAL
Window,											
Trim,											
Finish,											
Caulking,											
Grille,											
Drip cap,											
TOTAL											

Sliding Window System

SYSTEM DESCRIPTION	UNITS	QUAN.	LABOR-HOURS	COST EACH			EXTENSIONS				EACH
				MAT.	INST.	TOTAL	QUAN.	HOURS	MAT.	INST.	TOTAL
Window,											
Trim,											
Finish,											
Caulking,											
Grille,											
Drip cap,											
TOTAL											

Bow/Bay Window System

SYSTEM DESCRIPTION	UNITS	QUAN.	LABOR-HOURS	COST EACH			EXTENSIONS		EACH		
				MAT.	INST.	TOTAL	QUAN.	HOURS	MAT.	INST.	TOTAL
Window,											
Trim,											
Finish,											
Caulking,											
Grille,											
Drip cap,											
TOTAL											

Fixed Window System

SYSTEM DESCRIPTION	UNITS	QUAN.	LABOR-HOURS	COST EACH MAT.	COST EACH INST.	COST EACH TOTAL	EXTENSIONS QUAN.	EXTENSIONS HOURS	EACH MAT.	EACH INST.	EACH TOTAL
Window,											
Trim,											
Finish,											
Caulking,											
Grille,											
Drip cap,											
TOTAL											

Entrance Door System

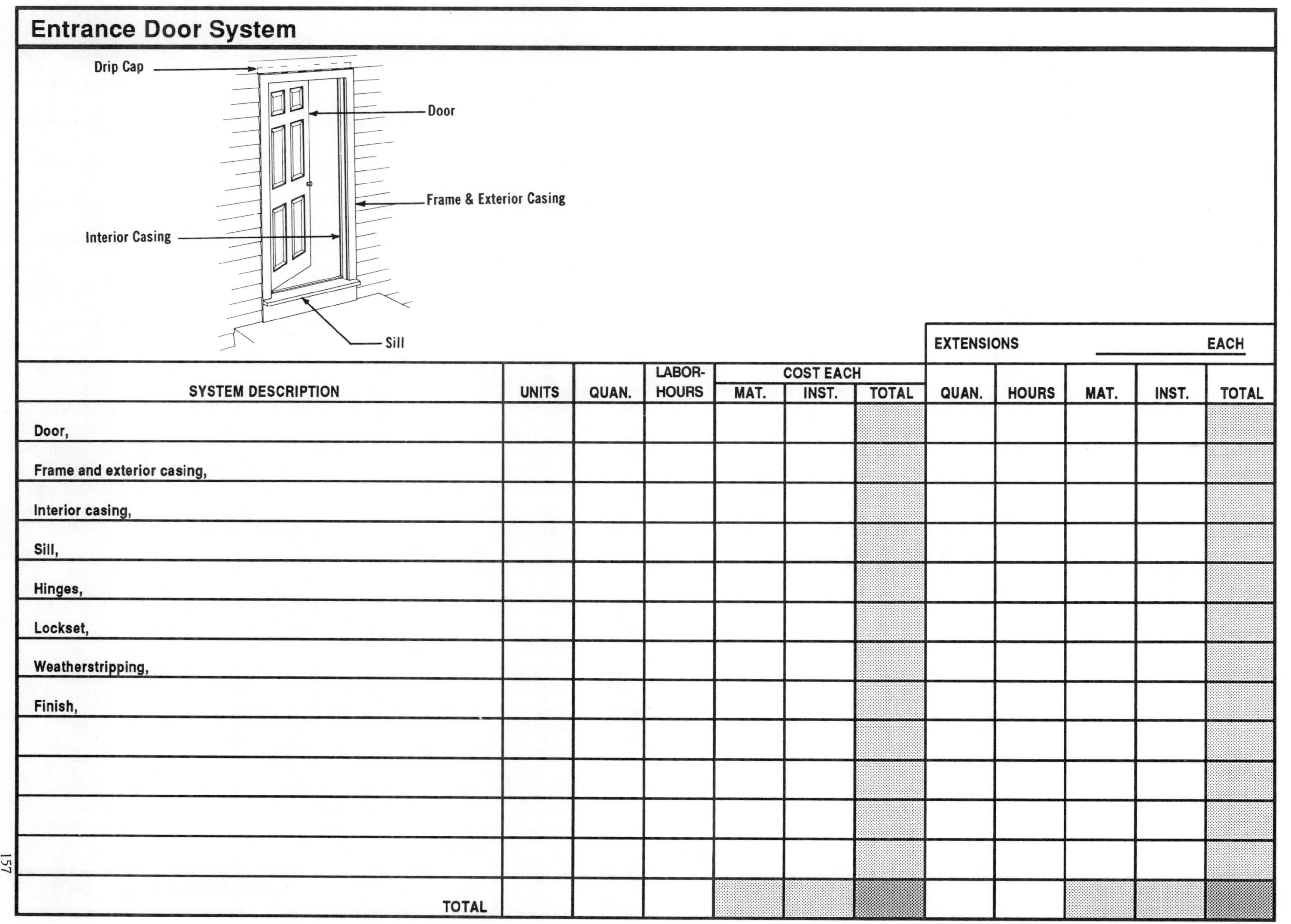

SYSTEM DESCRIPTION	UNITS	QUAN.	LABOR-HOURS	COST EACH			EXTENSIONS				EACH
				MAT.	INST.	TOTAL	QUAN.	HOURS	MAT.	INST.	TOTAL
Door,											
Frame and exterior casing,											
Interior casing,											
Sill,											
Hinges,											
Lockset,											
Weatherstripping,											
Finish,											
TOTAL											

Sliding Door System

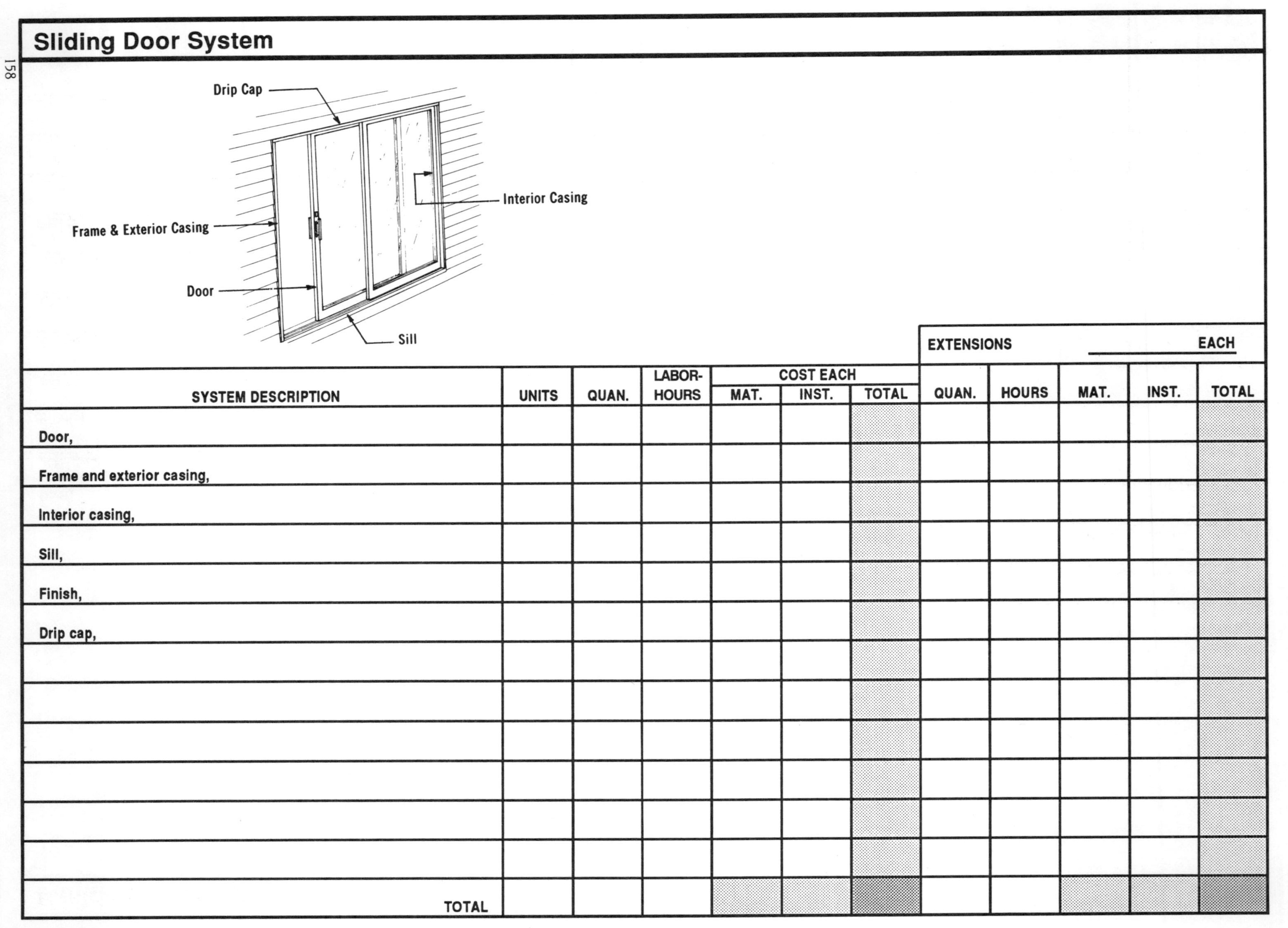

SYSTEM DESCRIPTION	UNITS	QUAN.	LABOR-HOURS	COST EACH			EXTENSIONS				EACH
				MAT.	INST.	TOTAL	QUAN.	HOURS	MAT.	INST.	TOTAL
Door,											
Frame and exterior casing,											
Interior casing,											
Sill,											
Finish,											
Drip cap,											
TOTAL											

Residential Garage Door System

| SYSTEM DESCRIPTION | UNITS | QUAN. | LABOR-HOURS | COST EACH | | | EXTENSIONS | | | | EACH |
				MAT.	INST.	TOTAL	QUAN.	HOURS	MAT.	INST.	TOTAL
Door,											
Jamb and header blocking,											
Exterior trim,											
Finish,											
Weatherstripping,											
Drip cap,											
TOTAL											

Aluminum Window System

| SYSTEM DESCRIPTION | UNITS | QUAN. | LABOR-HOURS | COST EACH | | | EXTENSIONS | | | | EACH |
				MAT.	INST.	TOTAL	QUAN.	HOURS	MAT.	INST.	TOTAL
Window,											
Blocking,											
Drywall,											
Corner bead,											
Finish drywall,											
Sill,											
TOTAL											

Gable End Roofing System

| SYSTEM DESCRIPTION | UNITS | QUAN. | LABOR-HOURS | COST PER S.F. | | | EXTENSIONS | | | | S.F. |
				MAT.	INST.	TOTAL	QUAN.	HOURS	MAT.	INST.	TOTAL
Shingles,											
Drip edge,											
Building paper,											
Ridge shingles,											
Soffit and fascia,											
Rake trim,											
Gutter,											
Downspouts,											
TOTAL											

Hip Roof Roofing System

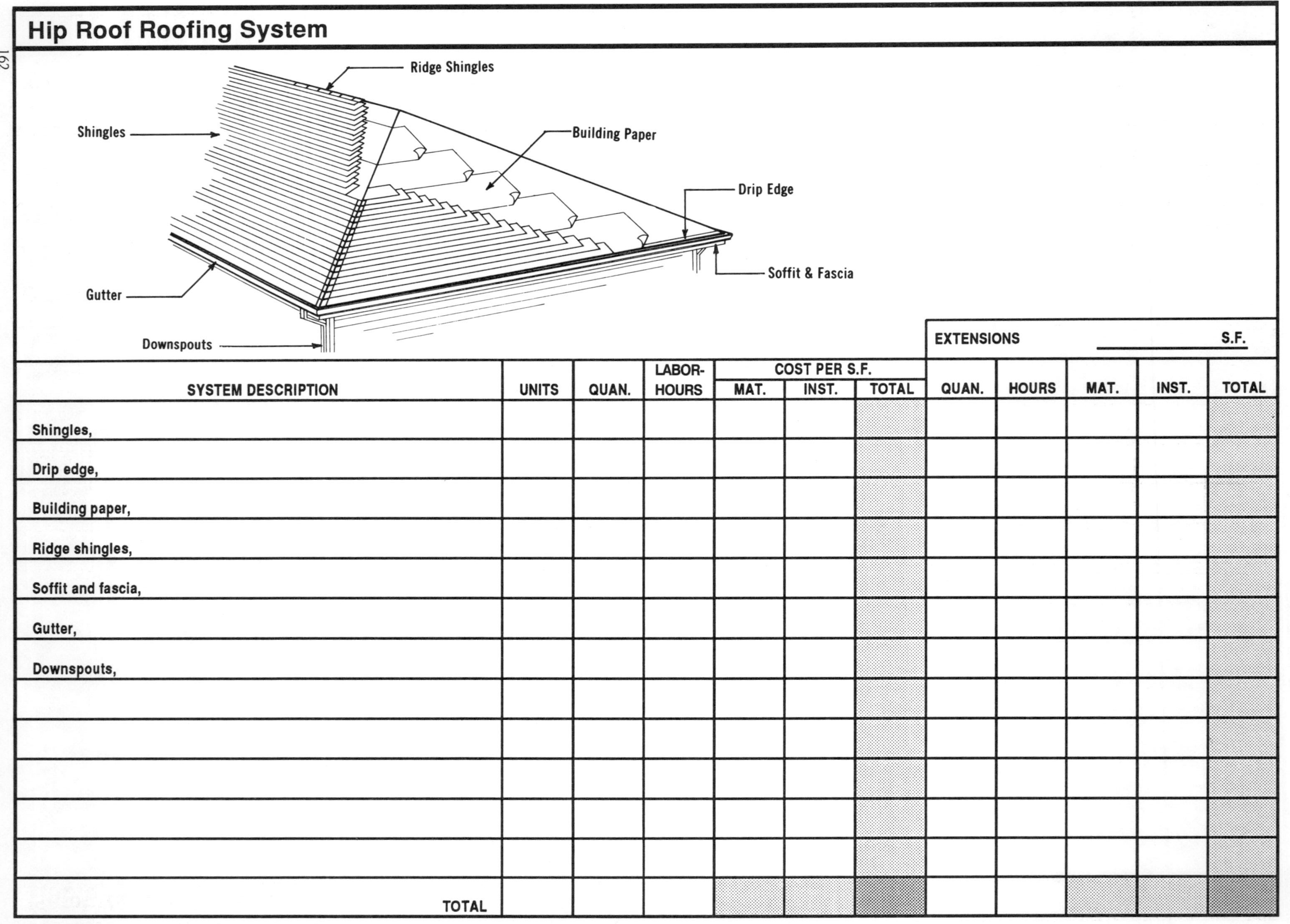

SYSTEM DESCRIPTION	UNITS	QUAN.	LABOR-HOURS	COST PER S.F.			EXTENSIONS				S.F.
				MAT.	INST.	TOTAL	QUAN.	HOURS	MAT.	INST.	TOTAL
Shingles,											
Drip edge,											
Building paper,											
Ridge shingles,											
Soffit and fascia,											
Gutter,											
Downspouts,											
TOTAL											

Gambrel Roofing System

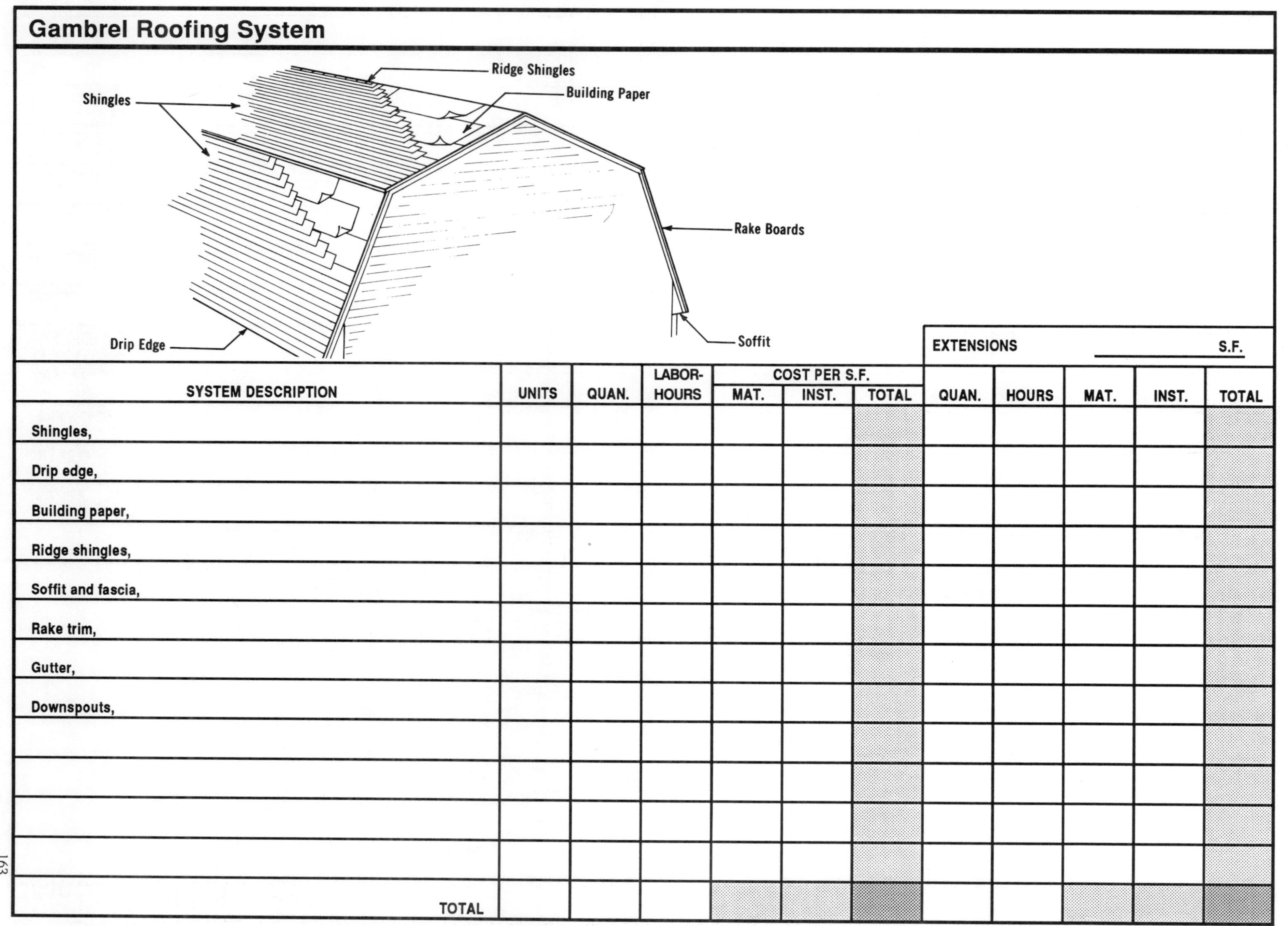

SYSTEM DESCRIPTION	UNITS	QUAN.	LABOR-HOURS	COST PER S.F.			EXTENSIONS				
				MAT.	INST.	TOTAL	QUAN.	HOURS	MAT.	INST.	TOTAL
Shingles,											
Drip edge,											
Building paper,											
Ridge shingles,											
Soffit and fascia,											
Rake trim,											
Gutter,											
Downspouts,											
TOTAL											

EXTENSIONS _______ S.F.

Mansard Roofing System

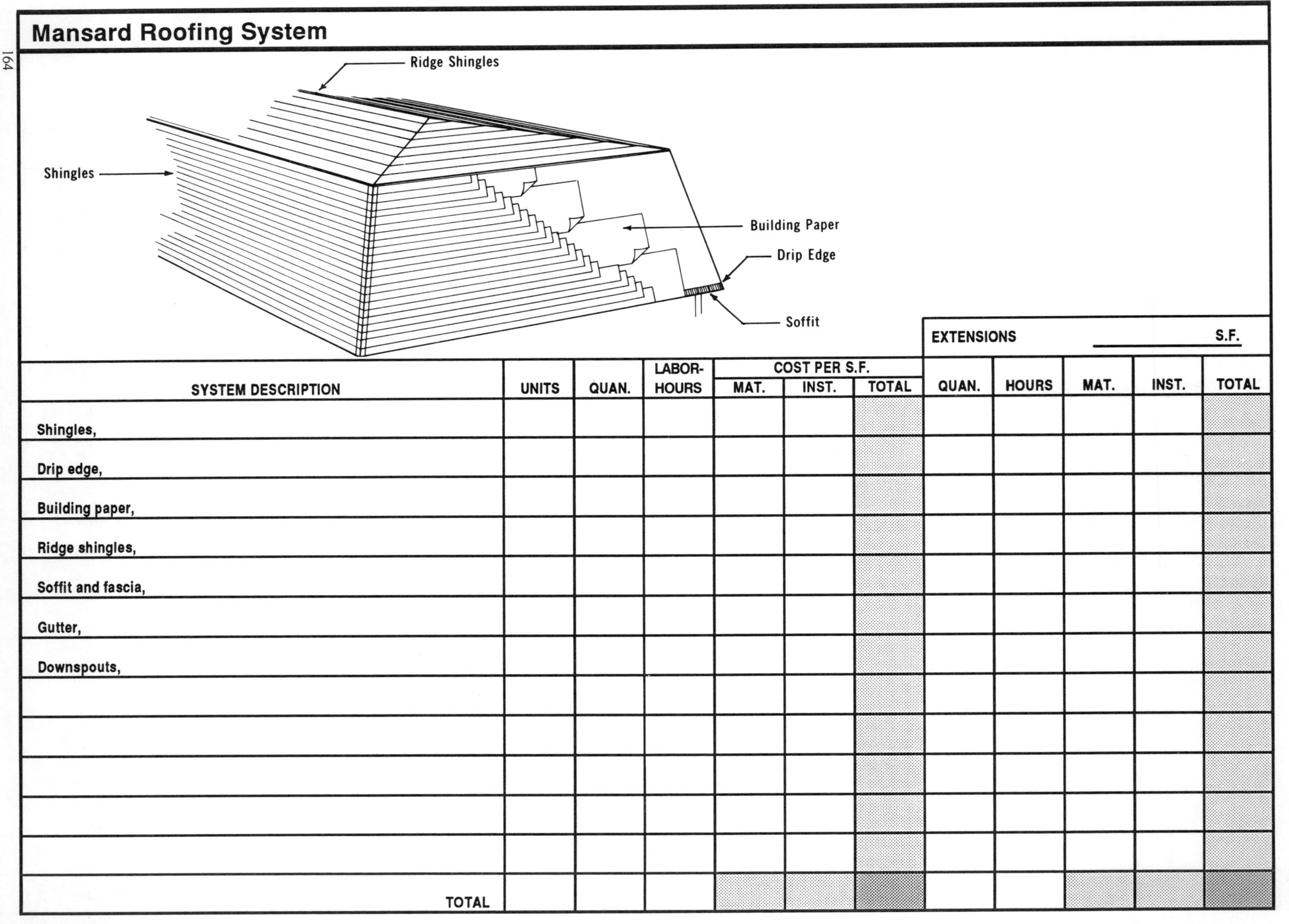

SYSTEM DESCRIPTION	UNITS	QUAN.	LABOR-HOURS	COST PER S.F.			EXTENSIONS				S.F.
				MAT.	INST.	TOTAL	QUAN.	HOURS	MAT.	INST.	TOTAL
Shingles,											
Drip edge,											
Building paper,											
Ridge shingles,											
Soffit and fascia,											
Gutter,											
Downspouts,											
TOTAL											

Shed Roofing System

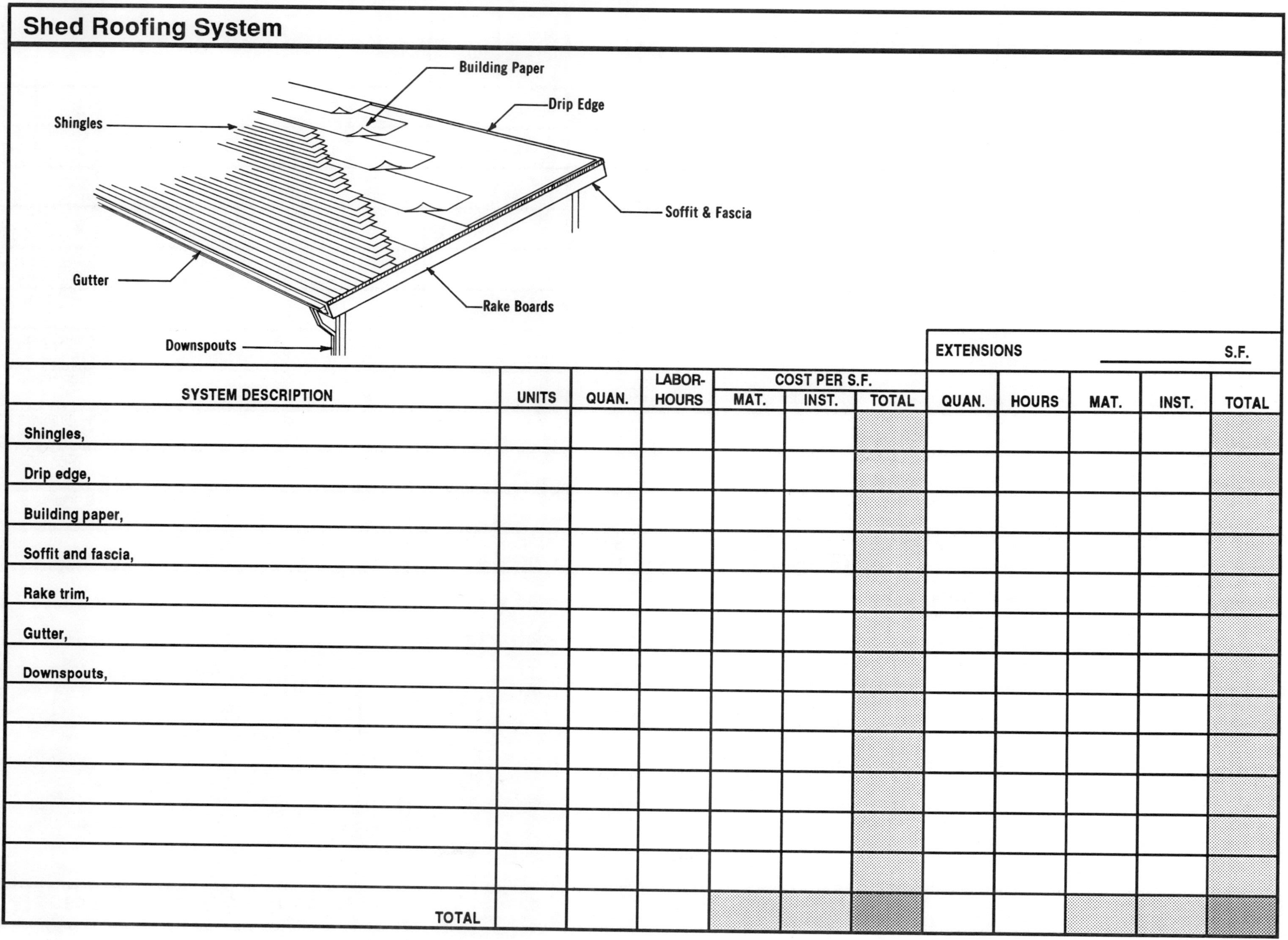

SYSTEM DESCRIPTION	UNITS	QUAN.	LABOR-HOURS	COST PER S.F. MAT.	COST PER S.F. INST.	COST PER S.F. TOTAL	EXTENSIONS QUAN.	EXTENSIONS HOURS	EXTENSIONS MAT.	EXTENSIONS INST.	EXTENSIONS TOTAL ___ S.F.
Shingles,											
Drip edge,											
Building paper,											
Soffit and fascia,											
Rake trim,											
Gutter,											
Downspouts,											
TOTAL											

Gable Dormer Roofing System

SYSTEM DESCRIPTION	UNITS	QUAN.	LABOR-HOURS	COST PER S.F.			EXTENSIONS QUAN.	HOURS	MAT.	INST.	TOTAL
				MAT.	INST.	TOTAL					S.F.
Shingles,											
Drip edge,											
Building paper,											
Ridge shingles,											
Soffit and fascia,											
Flashing,											
TOTAL											

Shed Dormer Roofing System

| SYSTEM DESCRIPTION | UNITS | QUAN. | LABOR-HOURS | COST PER S.F. | | | EXTENSIONS QUAN. | HOURS | MAT. | INST. | TOTAL |
				MAT.	INST.	TOTAL					S.F.
Shingles,											
Drip edge,											
Building paper,											
Soffit and fascia,											
Flashing,											
TOTAL											

Skylight/Skywindow System

SYSTEM DESCRIPTION	UNITS	QUAN.	LABOR-HOURS	COST EACH			EXTENSIONS		EACH		
				MAT.	INST.	TOTAL	QUAN.	HOURS	MAT.	INST.	TOTAL
Skylight or skywindow,											
Trimmer rafters,											
Headers,											
Curb,											
Flashing,											
Interior trim,											
TOTAL											

Drywall and Thincoat Wall System

| SYSTEM DESCRIPTION | UNITS | QUAN. | LABOR-HOURS | COST PER S.F. | | | EXTENSIONS | | | | S.F. |
				MAT.	INST.	TOTAL	QUAN.	HOURS	MAT.	INST.	TOTAL
Drywall,											
Finish,											
Corners,											
Painting,											
Trim,											
TOTAL											

Drywall and Thincoat Ceiling System

EXTENSIONS _______ S.F.

SYSTEM DESCRIPTION	UNITS	QUAN.	LABOR-HOURS	COST PER S.F.			QUAN.	HOURS	MAT.	INST.	TOTAL
				MAT.	INST.	TOTAL					
Drywall,											
Finish,											
Corners,											
Painting,											
TOTAL											

Plaster and Stucco Wall System

| SYSTEM DESCRIPTION | UNITS | QUAN. | LABOR-HOURS | COST PER S.F. | | | EXTENSIONS QUAN. | HOURS | MAT. | INST. | TOTAL |
				MAT.	INST.	TOTAL					S.F.
Plaster,											
Lath,											
Corners,											
Painting,											
Trim,											
TOTAL											

Plaster and Stucco Ceiling System

SYSTEM DESCRIPTION	UNITS	QUAN.	LABOR-HOURS	COST PER S.F.			EXTENSIONS QUAN.	HOURS	MAT.	INST.	TOTAL
				MAT.	INST.	TOTAL					
Plaster,											
Lath,											
Corners,											
Painting,											
TOTAL											

EXTENSIONS ___________ S.F.

Interior Door System

SYSTEM DESCRIPTION	UNITS	QUAN.	LABOR-HOURS	COST EACH			EXTENSIONS				EACH
				MAT.	INST.	TOTAL	QUAN.	HOURS	MAT.	INST.	TOTAL
Door,											
Frame,											
Trim,											
Hinges,											
Lockset,											
Finish,											
TOTAL											

Closet Door System

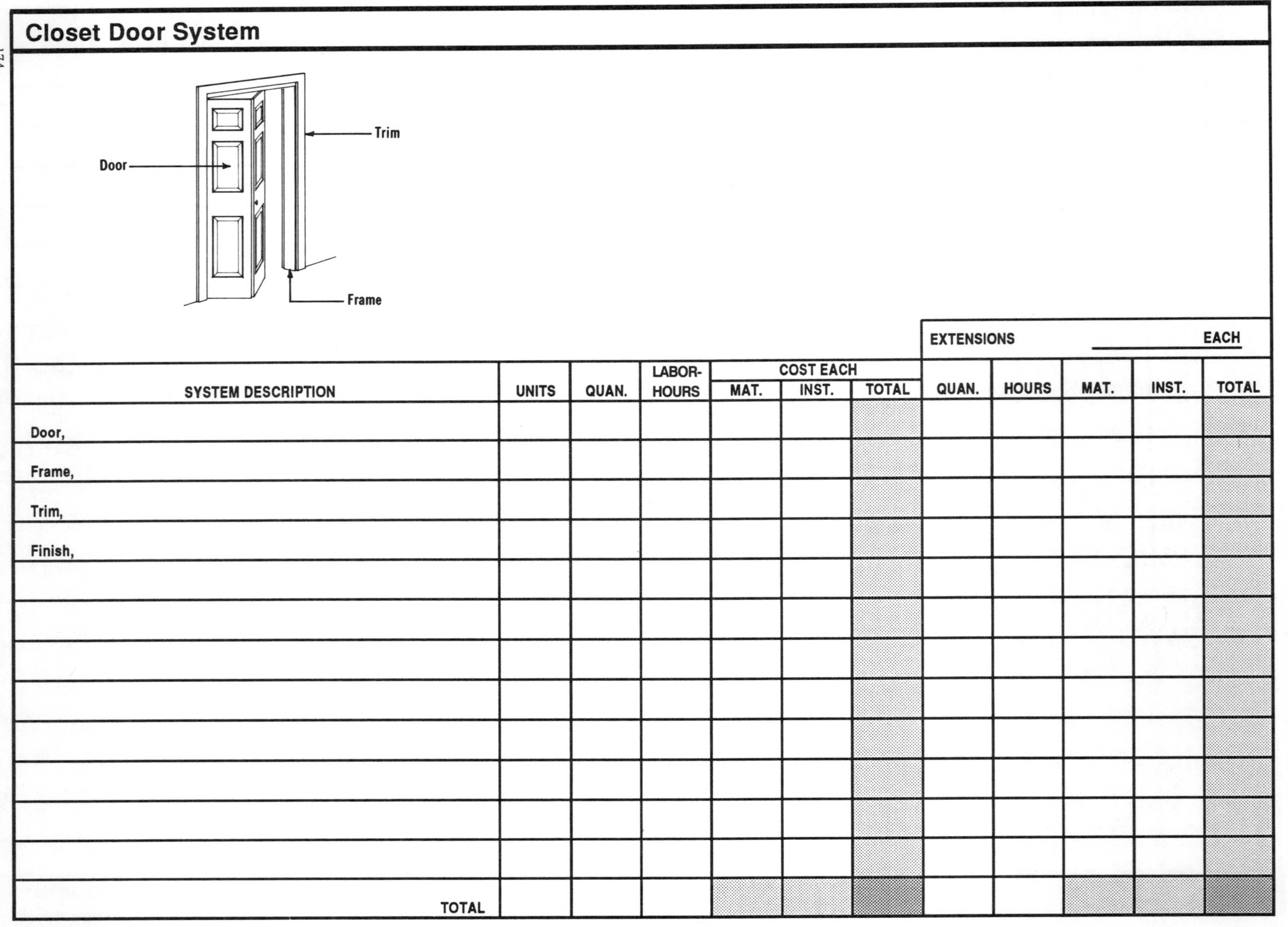

SYSTEM DESCRIPTION	UNITS	QUAN.	LABOR-HOURS	COST EACH			EXTENSIONS				EACH
				MAT.	INST.	TOTAL	QUAN.	HOURS	MAT.	INST.	TOTAL
Door,											
Frame,											
Trim,											
Finish,											
TOTAL											

Section Five

Reference Aids

This section consists of reference tables and explanations that can be used to develop stock lists and calculate material quantities. The tables contain a wealth of statistical information which you can use to help make your construction projects more economical and profitable. Also included is a list of the abbreviations used in this book.

Number of Board Feet in Various Sized Pieces of Lumber

This table shows how many board feet are in various pieces of lumber for the lengths indicated.

Size (in inches)								

<table>
<tr><td rowspan="2">Size (in inches)</td><td colspan="8" align="center">Length of Piece</td></tr>
<tr><td>8'</td><td>10'</td><td>12'</td><td>14'</td><td>16'</td><td>18'</td><td>20'</td><td>22'</td></tr>
<tr><td>2 x 3</td><td>4</td><td>5</td><td>6</td><td>7</td><td>8</td><td>9</td><td>10</td><td>11</td></tr>
<tr><td>2 x 4</td><td>5-1/3</td><td>6-2/3</td><td>8</td><td>9-1/3</td><td>10-2/3</td><td>12</td><td>13-1/3</td><td>14-2/3</td></tr>
<tr><td>2 x 6</td><td>8</td><td>10</td><td>12</td><td>14</td><td>16</td><td>18</td><td>20</td><td>22</td></tr>
<tr><td>2 x 8</td><td>10-2/3</td><td>13-1/3</td><td>16</td><td>18-2/3</td><td>21-1/3</td><td>24</td><td>26-2/3</td><td>29-1/3</td></tr>
<tr><td>2 x 12</td><td>16</td><td>20</td><td>24</td><td>28</td><td>32</td><td>36</td><td>40</td><td>44</td></tr>
<tr><td>3 x 4</td><td>8</td><td>10</td><td>12</td><td>14</td><td>16</td><td>18</td><td>20</td><td>22</td></tr>
<tr><td>3 x 6</td><td>12</td><td>15</td><td>18</td><td>21</td><td>24</td><td>27</td><td>30</td><td>33</td></tr>
<tr><td>3 x 8</td><td>16</td><td>20</td><td>24</td><td>28</td><td>32</td><td>36</td><td>40</td><td>44</td></tr>
<tr><td>3 x 10</td><td>20</td><td>25</td><td>30</td><td>35</td><td>40</td><td>45</td><td>50</td><td>55</td></tr>
<tr><td>3 x 12</td><td>24</td><td>30</td><td>36</td><td>42</td><td>48</td><td>54</td><td>60</td><td>66</td></tr>
<tr><td>4 x 4</td><td>10-2/3</td><td>13-1/3</td><td>16</td><td>18-2/3</td><td>21-1/3</td><td>24</td><td>26-2/3</td><td>29-1/3</td></tr>
<tr><td>4 x 6</td><td>16</td><td>20</td><td>24</td><td>28</td><td>32</td><td>36</td><td>40</td><td>44</td></tr>
<tr><td>4 x 8</td><td>21-1/3</td><td>26-2/3</td><td>32</td><td>37-1/3</td><td>42-2/3</td><td>48</td><td>53-1/3</td><td>58-2/3</td></tr>
<tr><td>6 x 6</td><td>24</td><td>30</td><td>36</td><td>42</td><td>48</td><td>54</td><td>60</td><td>66</td></tr>
<tr><td>6 x 8</td><td>32</td><td>40</td><td>48</td><td>56</td><td>64</td><td>72</td><td>80</td><td>88</td></tr>
<tr><td>6 x 10</td><td>40</td><td>50</td><td>60</td><td>70</td><td>80</td><td>90</td><td>100</td><td>110</td></tr>
<tr><td>8 x 8</td><td>42-2/3</td><td>53-1/3</td><td>64</td><td>74-2/3</td><td>85-1/3</td><td>96</td><td>106-2/3</td><td>117-1/3</td></tr>
<tr><td>8 x 10</td><td>53-1/3</td><td>66-2/3</td><td>80</td><td>93-1/3</td><td>106-2/3</td><td>120</td><td>133-1/3</td><td>146-2/3</td></tr>
<tr><td>8 x 12</td><td>64</td><td>80</td><td>96</td><td>112</td><td>128</td><td>144</td><td>160</td><td>176</td></tr>
<tr><td>10 x 10</td><td>66-2/3</td><td>83-1/3</td><td>100</td><td>116-2/3</td><td>133-1/3</td><td>150</td><td>166-2/3</td><td>183-1/3</td></tr>
<tr><td>10 x 12</td><td>80</td><td>100</td><td>120</td><td>140</td><td>160</td><td>180</td><td>200</td><td>220</td></tr>
<tr><td>12 x 14</td><td>112</td><td>140</td><td>168</td><td>196</td><td>224</td><td>252</td><td>280</td><td>308</td></tr>
</table>

Board Foot Multipliers

This table is used for computing the number of board feet in any length of dimension lumber. Find the size of lumber to be used, and read across for the multiplier. For example, to find out how many board feet of lumber are in an 8' length of 2 x 4 lumber, look for the multiplier for 2 x 4's (0.667) and multiply it by the length of lumber (8'). The result is 0.667 x 8, or 5.334 board feet.

Nominal Size (in inches)	Multiply Length by	Nominal Size (in inches)	Multiply Length by
2 x 2	0.333	4 x 4	1.333
2 x 3	0.500	4 x 6	2.000
2 x 4	0.667	4 x 8	2.667
2 x 6	1.000	4 x 10	3.333
2 x 8	1.333	4 x 12	4.000
2 x 10	1.667		
2 x 12	2.000	6 x 6	3.000
		6 x 8	4.000
3 x 3	0.750	6 x 10	5.000
3 x 4	1.000	6 x 12	6.000
3 x 6	1.500		
3 x 8	2.000	8 x 8	5.333
3 x 10	2.500	8 x 10	6.667
3 x 12	3.000	8 x 12	8.000

Factors for the Board Foot
Measure of Floor or Ceiling Joists

This table allows you to compute the amount of lumber (in board feet) for the floor or ceiling joists of a known area, based on the size of lumber and the spacing. To use this chart, simply multiply the area of the room or building in question by the factor across from the appropriate board size and spacing.

Ceiling Joists			
Joist Size	Inches On Center	Board Feet per Square Foot of Area	Approximate Nails Lbs. per MBM
2″ x 4″	12″	.78	17
	16″	.59	19
	20″	.48	19
2″ x 6″	12″	1.15	11
	16″	.88	13
	20″	.72	13
	24″	.63	13
2″ x 8″	12″	1.53	9
	16″	1.17	9
	20″	.96	9
	24″	.84	9
2″ x 10″	12″	1.94	7
	16″	1.47	7
	20″	1.21	7
	24″	1.04	7
3″ x 8″	12″	2.32	6
	16″	1.76	6
	20″	1.44	6
	24″	1.25	6

EXAMPLE:

You have a 10′ x 20′ area that you have to use 2″ x 6″ floor joists. To calculate the number of board feet you will enter the 2″ x 6″ section of the table. If your spacing is to be 16″ O.C. your calculations will be:

$$10' \times 20' \times .88 \ \frac{B.F.}{S.F.} = 176 \ B.F.$$

You will also need approximately

$$176 \ B.F. \times \frac{13 \ Lbs.}{100 \ B.F.} = 2 \ 1/4 \ Lbs. \ of \ nails$$

Pieces of Studding Required
for Walls, Floors, or Ceilings

This table lists the number of pieces of studding or furring needed for framing walls, floors, ceilings, partitions, etc. based on the length of the item being framed and its required spacing. Note that no provisions have been made for the doubling of studs at corners, window openings, door openings, etc. These must be added as called for.

Length of Wall, Floor or Ceiling (in feet)	On Center Spacing			
	12"	16"	20"	24"
8	9	7	6	5
9	10	8	6	6
10	11	9	7	6
11	12	9	8	7
12	13	10	8	7
13	14	11	9	8
14	15	12	9	8
15	16	12	10	9
16	17	13	11	9
17	18	14	11	10
18	19	15	12	10
19	20	15	12	11
20	21	16	13	11
21	22	17	14	12
22	23	18	14	12
23	24	18	15	13
24	25	19	15	13
25	26	20	16	14
26	27	21	17	14
27	28	21	17	15
28	29	22	18	15
29	30	23	18	16
30	31	24	19	16
32	33	25	20	17
34	35	27	22	18
36	37	28	23	19

EXAMPLE:
You have to frame a partition 18' long 10' high with 2" x 4" lumber at 16" spacing.

You will need 15 pieces of 10' long 2" x 4"'s. You should then add for blocking/bridging etc.

(excerpted from *How to Estimate Building Losses and Construction Costs*, Paul I. Thomas, Prentice-Hall)

Multipliers for Computing
Required Studding or Furring

This table allows you to compute the number of pieces of studding or furring required for framing walls, floors, roofs, partitions, etc.

Spacing (in inches)	Factor by Which to Multiply the Width of Room, Length of Wall or Roof
12	1.00
16	.75
18	.67 Add one for end in each case.
20	.60
24	.50

Examples:	Answers:
A room 16' wide requires studs 16" O.C.	16' x .75 + 1 = 13 joists
A wall 30' long requires studs 18" O.C.	30' x .67 + 1 = 21 studs
A gable roof 40' long requires studs 24" O.C.	40' x .50 + 1 = 21 rafters

(excerpted from *How to Estimate Building Losses and Construction Costs*, Paul I. Thomas, Prentice Hall)

Partition Framing

This table can be used to compute the board feet of lumber required for each square foot of wall area to be framed, based on the lumber size and spacing design.

Stud Size	Studs Including Sole and Cap Plates			Horizontal Bracing in All Partitions		Horizontal Bracing In Bearing Partitions Only	
	Inches On Center	Board Feet per Square Foot of Partition Area	Lbs. Nails per MBM of Stud Framing	Board Feet per Square Foot of Partition Area	Lbs. Nails per MBM of Bracing	Board Feet per Square Foot of Partition Area	Lbs. Nails per MBM of Bracing
2" x 3"	12"	.91	25	.04	145	.01	145
	16"	.83	25	.04	111	.01	111
	20"	.78	25	.04	90	.01	90
	24"	.76	25	.04	79	.01	79
2" x 4"	12"	1.22	19	.05	108	.02	108
	16"	1.12	19	.05	87	.02	87
	20"	1.05	19	.05	72	.02	72
	24"	1.02	19	.05	64	.02	64
2" x 6"	16"	1.38	19			.04	59
	20"	1.29	16			.04	48
	24"	1.22	16			.04	43
2" x 4" Staggered	8"	1.69	22				
3" x 4"	16"	1.35	17				
2" x 4" 2" Way	16"	1.08	19				

Exterior Wall Stud Framing

This table allows you to compute the board feet of lumber required for each square foot of exterior wall to be framed based on the lumber size and spacing design.

Stud Size	Inches On Center	Studs Including Corner Bracing		Horizontal Bracing Midway Between Plates	
		Board Feet per Square Foot of Ext. Wall Area	Lbs. of Nails per MBM of Stud Framing	Board Feet per Square Foot of Ext. Wall Area	Lbs. of Nails per MBM of Bracing
2" x 3"	16"	.78	30	.03	117
	20"	.74	30	.03	97
	24"	.71	30	.03	85
2" x 4"	16"	1.05	22	.04	87
	20"	.98	22	.04	72
	24"	.94	22	.04	64
2" x 6"	16"	1.51	15	.06	59
	20"	1.44	15	.06	48
	24"	1.38	15	.06	43

Floor Framing

This table can be used to compute the board feet required for each square foot of floor area based on the size of lumber to be used and the spacing design.

		Floor Joists			Block Over Main Bearing	
Joist Size	Inches On Center	Board Feet per Square Foot of Floor Area	Nails Lbs. per MBM		Board Feet per Square Foot of Floor Area	Nails Lbs. per MBM Blocking
2" x 6"	12"	1.28	10		.16	133
	16"	1.02	10		.03	95
	20"	.88	10		.03	77
	24"	.78	10		.03	57
2" x 8"	12"	1.71	8		.04	100
	16"	1.36	8		.04	72
	20"	1.17	8		.04	57
	24"	1.03	8		.05	43
2" x 10"	12"	2.14	6		.05	70
	16"	1.71	6		.05	57
	20"	1.48	6		.06	46
	24"	1.30	6		.06	34
2" x 12"	12"	2.56	5		.06	66
	16"	2.05	5		.06	47
	20"	1.77	5		.07	39
	24"	1.56	5		.07	29
3" x 8"	12"	2.56	5		.04	39
	16"	2.05	5		.05	57
	20"	1.77	5		.06	45
	24"	1.56	5		.06	33
3" x 10"	12"	3.20	4		.05	72
	16"	2.56	4		.07	46
	20"	2.21	4		.07	36
	24"	1.95	4		.08	26

Furring Quantities

This table provides multiplication factors for converting square feet of wall area requiring furring into board feet of lumber. These figures are based on lumber size and spacing.

	Board Feet per Square Feet of Wall Area				Lbs. Nails per MBM of Furring
	Spacing Center to Center				
Size	12"	16"	20"	24"	
1" x 2"	.18	.14	.11	.10	55
1" x 3"	.28	.21	.17	.14	37

Conversion Factors for Wood Joists and Rafters

This table is used to compute actual lengths or board feet of lumber for roofs of various inclines. To compute actual rafter quantities for incline roofs, multiply the quantities figured for a flat roof by the factors as shown in the table. Please note that this table DOES NOT INCLUDE QUANTITIES FOR CANTILEVERED OVERHANGS.

Roof Slope	Approximate Angle	Factor	Roof Slope	Approximate Angle	Factor
Flat	0°	1.000	12 in 12	45.0°	1.414
1 in 12	4.8°	1.003	13 in 12	47.3°	1.474
2 in 12	9.5°	1.014	14 in 12	49.4°	1.537
3 in 12	14.0°	1.031	15 in 12	51.3°	1.601
4 in 12	18.4°	1.054	16 in 12	53.1°	1.667
5 in 12	22.6°	1.083	17 in 12	54.8°	1.734
6 in 12	26.6°	1.118	18 in 12	56.3°	1.803
7 in 12	30.3°	1.158	19 in 12	57.7°	1.873
8 in 12	33.7°	1.202	20 in 12	59.0°	1.943
9 in 12	36.9°	1.250	21 in 12	60.3°	2.015
10 in 12	39.8°	1.302	22 in 12	61.4°	2.088
11 in 12	42.5°	1.357	23 in 12	62.4°	2.162

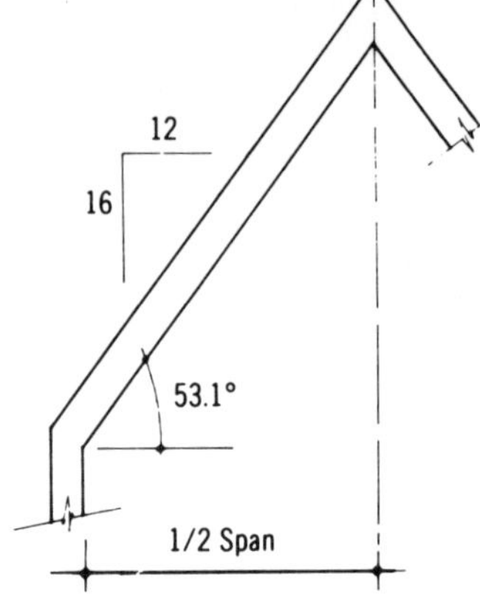

Inclined to Horizontal Sections

Roofing Factors When Rate of Rise is Known

No. Inches Rise per Foot of Run	*Pitch	Rafter Length in Inches per Foot of Run	To Obtain Rafter Length Multiply Run (in feet) By:
3.0	1/8	12.37 ÷ 12 =	1.030
3.5			1.045
4.0	1/6	12.65	1.060
4.5			1.075
5.0	5/24	13.00	1.090
5.5			1.105
6.0	1/4	13.42	1.120
6.5			1.140
7.0	7/24	13.89	1.160
7.5			1.185
8.0	1/3	14.42	1.201
8.5			1.230
9.0	3/8	15.00	1.250
9.5			1.280
10.0	5/12	16.62	1.301
10.5			1.335
11.0	11/24	16.28	1.360
11.5			1.390
12.0	1/2	16.97	1.420

*Pitch is determined by dividing the inches rise-per-foot-of-run by 24. (Most roofs of dwellings are constructed with a pitch of 1/8, 1/6, 1/4, 1/3 or 1/2.)
(excerpted from *How to Estimate Building Losses and Construction Costs*, Paul I. Thomas, Prentice Hall)

Flat Roof Framing

This table can be used to compute the amount of lumber in board feet required to frame a flat roof, based on the size lumber required.

Flat Roof Framing			
Joist Size	Inches On Center	Board Feet per Square Foot of Ceiling Area	Nails Lbs. per MBM
2" x 6"	12"	1.17	10
	16"	.91	10
	20"	.76	10
	24"	.65	10
2" x 8"	12"	1.56	8
	16"	1.21	8
	20"	1.01	8
	24"	.86	8
2" x 10"	12"	1.96	6
	16"	1.51	6
	20"	1.27	6
	24"	1.08	6
2" x 12"	12"	2.35	5
	16"	1.82	5
	20"	1.52	5
	24"	1.30	5
3" x 8"	12"	2.35	5
	16"	1.82	5
	20"	1.52	5
	24"	1.30	5
3" x 10"	12"	2.94	4
	16"	2.27	4
	20"	1.90	4
	24"	1.62	4

MBM; MFBM = Thousand Feet Board Measure

Allowance Factors for Roof Overhangs

Horizontal Span	Roof Overhang Measured Horizontally							
	0'-6"	1'-0"	1'-6"	2'-0"	2'-6"	3'-0"	3'-6"	4'-0"
6'	1.083	1.167	1.250	1.333	1.417	1.500	1.583	1.667
7'	1.071	1.143	1.214	1.286	1.357	1.429	1.500	1.571
8'	1.063	1.125	1.188	1.250	1.313	1.375	1.438	1.500
9'	1.056	1.111	1.167	1.222	1.278	1.333	1.389	1.444
10'	1.050	1.100	1.150	1.200	1.250	1.300	1.350	1.400
11'	1.045	1.091	1.136	1.182	1.227	1.273	1.318	1.364
12'	1.042	1.083	1.125	1.167	1.208	1.250	1.292	1.333
13'	1.038	1.077	1.115	1.154	1.192	1.231	1.269	1.308
14'	1.036	1.071	1.107	1.143	1.179	1.214	1.250	1.286
15'	1.033	1.067	1.100	1.133	1.167	1.200	1.233	1.267
16'	1.031	1.063	1.094	1.125	1.156	1.188	1.219	1.250
17'	1.029	1.059	1.088	1.118	1.147	1.176	1.206	1.235
18'	1.028	1.056	1.083	1.111	1.139	1.167	1.194	1.222
19'	1.026	1.053	1.079	1.105	1.132	1.158	1.184	1.211
20'	1.025	1.050	1.075	1.100	1.125	1.150	1.175	1.200
21'	1.024	1.048	1.071	1.095	1.119	1.143	1.167	1.190
22'	1.023	1.045	1.068	1.091	1.114	1.136	1.159	1.182
23'	1.022	1.043	1.065	1.087	1.109	1.130	1.152	1.174
24'	1.021	1.042	1.063	1.083	1.104	1.125	1.146	1.167
25'	1.020	1.040	1.060	1.080	1.100	1.120	1.140	1.160
26'	1.019	1.038	1.058	1.077	1.096	1.115	1.135	1.154
27'	1.019	1.037	1.056	1.074	1.093	1.111	1.130	1.148
28'	1.018	1.036	1.054	1.071	1.089	1.107	1.125	1.143
29'	1.017	1.034	1.052	1.069	1.086	1.103	1.121	1.138
30'	1.017	1.033	1.050	1.067	1.083	1.100	1.117	1.133
32'	1.016	1.031	1.047	1.063	1.078	1.094	1.109	1.125

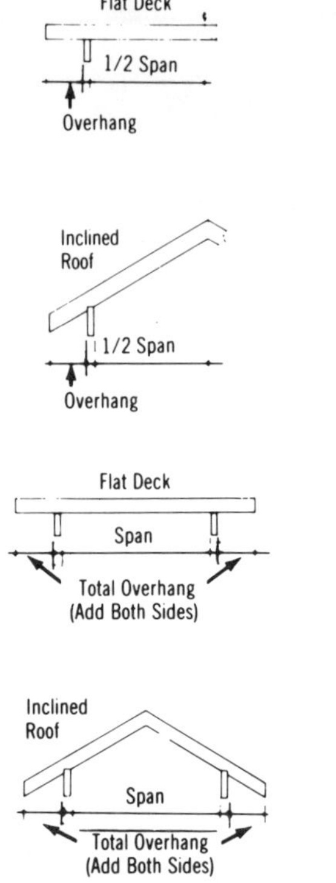

Overhang Sections

Pitched Roof Framing

This table is used to compute the board feet of lumber required per square foot of roof area, based on the required spacing and the lumber size to be used.

	Rafters Including Collar Ties, Hip and Valley Rafters, Ridge Poles							
	Spacing Center to Center							
	12"		16"		20"		24"	
Rafter Size	Board Feet per Square Foot of Roof Area	Nails Lbs. per MBM	Board Feet per Square Foot of Roof Area	Nails Lbs. per MBM	Board Feet per Square Foot of Roof Area	Nails Lbs. per MBM	Board Feet per Square Foot of Roof Area	Nails Lbs. per MBM
2" x 4"	.89	17	.71	17	.59	17	.53	17
2" x 6"	1.29	12	1.02	12	.85	12	.75	12
2" x 8"	1.71	9	1.34	9	1.12	9	.98	9
2" x 10"	2.12	7	1.66	7	1.38	7	1.21	7
2" x 12"	2.52	6	1.97	6	1.64	6	1.43	6
3" x 8"	2.52	6	1.97	6	1.64	6	1.43	6
3" x 10"	3.13	5	2.45	5	2.02	5	1.78	5

Material Required for On-the-Job Cut Bridging—Lineal Feet

Based on the size of the lumber used for joists, the total amount of lumber (in board feet) can be obtained for the various spacing of the joists using this table.

Size Joist in Inches	Spacing in Inches	Lineal Feet per Set (2)	Lineal Feet per Foot-of-Row
2 x 6	16	2.57	1.92
2 x 8	16	2.70	2.02
2 x 10	16	2.87	2.15
2 x 12	16	3.06	2.30
2 x 8	20	3.31	2.00
2 x 10	20	3.45	2.07
2 x 12	20	3.61	2.17
2 x 8	24	3.94	2.00
2 x 10	24	4.05	1.97
2 x 12	24	4.19	2.10

Note: Add to the total lineal feet developed from the table at least 10% cutting waste.

Example: A room 20 feet wide has two rows of bridging. The 2" x 12" joists are 16" on center.

2 x 20 L.F. x 2.30 ft. per foot-of-row = 92.0 L.F.

Add 10% waste 9.2

Total L.F. = 101.2

Round out to 101 L.F.

(Per set method would be, 30 sets x 3.06 = 91.8 to which must be added 10% waste.)

(excerpted from *How to Estimate Building Losses and Construction Costs*, Paul I. Thomas, Prentice Hall)

Board Feet Required for On-the-Job Cut Bridging

Based on the size of the lumber used for joists, the total lengths of lumber for various sized bridging can be obtained from this chart.

Cross Bridging—Board Feet per Square Foot of Floors, Ceiling or Flat Roof Area Nails—Pounds Per MBM of Bridging							
		1" x 3"		1" x 4"		2" x 3"	
Joist Size	Spacing	B.F.	Nails	B.F.	Nails	B.F.	Nails
2" x 8"	12"	.04	147	.05	112	.08	77
	16"	.04	120	.05	91	.08	61
	20"	.04	102	.05	77	.08	52
	24"	.04	83	.05	63	.08	42
2" x 10"	12"	.04	136	.05	103	.08	71
	16"	.04	114	.05	87	.08	58
	20"	.04	98	.05	74	.08	50
	24"	.04	80	.05	61	.08	41
2" x 12"	12"	.04	127	.05	96	.08	67
	16"	.04	108	.05	82	.08	55
	20"	.04	94	.05	71	.08	48
	24"	.04	78	.05	59	.08	39
3" x 8"	12"	.04	160	.05	122	.08	84
	16"	.04	127	.05	96	.08	66
	20"	.04	107	.05	81	.08	54
	24"	.04	86	.05	65	.08	44
3" x 10"	12"	.04	146	.05	111	.08	77
	16"	.04	120	.05	91	.08	62
	20"	.04	102	.05	78	.08	52
	24"	.04	83	.05	63	.08	42

Board Feet of Sheathing and Subflooring

This table is used to compute the board feet of sheathing or subflooring required per square foot of roof, ceiling or flooring.

Type	Size	Board Feet per Square Foot of Area	Diagonal			
			Lbs. Nails per MBM Lumber			
			Joist, Stud or Rafter Spacing			
			12"	16"	20"	24"
Surface 4 Sides	1" x 4"	1.22	58	46	39	32
(S4S)	1" x 6"	1.18	39	31	25	21
	1" x 8"	1.18	30	23	19	16
	1" x 10"	1.17	35	27	23	19
Tongue and Groove	1" x 4"	1.36	65	51	43	36
(T&G)	1" x 6"	1.26	42	33	27	23
	1" x 8"	1.22	31	24	20	17
	1" x 10"	1.20	36	28	24	19
Shiplap	1" x 4"	1.41	67	53	45	37
	1" x 6"	1.29	43	33	28	23
	1" x 8"	1.24	31	24	20	17
	1" x 10"	1.21	36	28	24	19

Wood Siding Factors

This table shows the factor by which area to be covered is multiplied to determine exact amount of surface material needed.

Item	Nominal Size	Width Overall	Face	Area Factor
Shiplap	1" x 6"	5-1/2"	5-1/8"	1.17
	1 x 8	7-1/4	6-7/8	1.16
	1 x 10	9-1/4	8-7/8	1.13
	1 x 12	11-1/4	10-7/8	1.10
Tongue and Groove	1 x 4	3-3/8	3-1/8	1.28
	1 x 6	5-3/8	5-1/8	1.17
	1 x 8	7-1/8	6-7/8	1.16
	1 x 10	9-1/8	8-7/8	1.13
	1 x 12	11-1/8	10-7/8	1.10
S4S	1 x 4	3-1/2	3-1/2	1.14
	1 x 6	5-1/2	5-1/2	1.09
	1 x 8	7-1/4	7-1/4	1.10
	1 x 10	9-1/4	9-1/4	1.08
	1 x 12	11-1/4	11-1/4	1.07
Solid paneling	1 x 6	5-7/16	5-7/16	1.19
	1 x 8	7-1/8	6-3/4	1.19
	1 x 10	9-1/8	8-3/4	1.14
	1 x 12	11-1/8	10-3/4	1.12
Bevel Siding*	1 x 4	3-1/2	3-1/2	1.60
	1 x 6	5-1/2	5-1/2	1.33
	1 x 8	7-1/4	7-1/4	1.28
	1 x 10	9-1/4	9-1/4	1.21
	1 x 12	11-1/4	11-1/4	1.17

Note: This area factor is strictly so-called milling waste. The cutting and fitting waste must be added.
*1" lap
(from *Western Wood Products Association*)

Milling and Cutting Waste Factors
for Wood Siding

This table lists the milling waste, the cutting waste, and the amount of nails required for various types of wood siding. Amounts are per 1,000 FBM of wood siding.

Type of Siding	Nominal Size in Inches	Lap in Inches 1" Lap	Pounds Nails per 1,000 FBM	Percentage of Waste
Bevel Siding	1 x 4	1	25-6d common	63
	1 x 6	1	25-6d common	35
	1 x 8	1-1/4	20-8d common	35
	1 x 10	1-1/2	20-8d common	30
Rustic and Drop Siding	1 x 4	Matched	40-8d common	33
	1 x 6	Matched	30-8d common	25
	1 x 8	Matched	25-8d common	20
Vertical Siding	1 x 6	Matched	25-8d finish	20
	1 x 8	Matched	20-8d finish	18
	1 x 10	Matched	20-8d finish	15
Batten Siding*	1 x 8	Rough	25	5
	1 x 10	Rough	20	5
	1 x 12	Rough	20	5
	1 x 8	Dressed	25	13
	1 x 10	Dressed	20	11
	1 x 12	Dressed	20	10
Plywood Siding	1/4	Sheets		5-10
	3/8	Sheets	15 per MSF	5-10
	5/8	Sheets		5-10

*For 1" x 10" boards, allow 1,334 lineal feet 1" x 2" joint strips for each 1,000 FBM of batten siding.
 Add 12 pounds 8d common nails.

(excerpted from *How to Estimate Building Losses and Construction Costs*, Paul I. Thomas, Prentice Hall)

Nails—Dimensions

This table describes the sizes, lengths, gauge number, and approximate number of nails one can expect per pound, based on the type of nail and use.

Penny Nail System					
Size	Length in inches	Gauge Number	Approximate Number to Pound		
			Common	Finishing	Casing
2d	1	15	850		
3d	1-1/4	14	550	640	
4d	1-1/2	12-1/2	350	456	
5d	1-3/4	12-1/2	230	328	
6d	2	11-1/2	180	273	228
7d	2-1/4	11	140	170	178
8d	2-1/2	10-1/4	100	151	133
9d	2-3/4	9-1/2	80	125	100
10d	3	9	65	107	96
12d	3-1/4	9	50		60
16d	3-1/2	8	40		50
20d	4	6	31		
30d	4-1/2	5	22		
40d	5	4	18		
50d	5-1/2	3	14		
60d	6	2	12		

*For 1" x 10" boards, allow 1,334 lineal feet 1" x 2" joint strips for each 1,000 FBM of battem siding.
 Add 12 pounds 8d common nails.

(excerpted from *How to Estimate Building Losses and Construction Costs*, Paul I. Thomas, Prentice Hall)

Size and Quantity of Nails for a Job

This table shows the size and approximate quantity of nails needed for various portions of a wood project.

Where a mix of nails is shown, judgment must be exercised as to the different sizes of lumber.

Kind of Framing and Size of Lumber	Size of Nails Used	Lbs. per 1,000 FBM
Sills and plates	10d, 16d & 20d	8
Wall and partition stud	10d & 16d	10
Joists and rafters		
2" x 6"	16d, some 20d	9
2" x 8"	16d, some 20d	8
2" x 10"	16d, some 20d	7
2" x 12"	16d, some 20d	6
Average for total house framing	8d, 10d, 16d, 20d	15
Wood cross bridging, 1" x 3"	8d	1 lb. per 12 sets
Furring (100 S.F.) on masonry	8d	1
Furring (100 S.F.) on studding	8d	1/2
Roof trusses	10d, 20d and 40d	10

Rough Hardware Allowances

Average Material Cost Allowances for Rough Hardware as a Percentage of Carpentry Material Costs	
Minimum	0.5%
Maximum	1.5%

Loose Fill Insulation

Square feet covered by a 40 lb. bag of mineral wool or glass fiber. (Includes area occupied by studding or joists.) Divide the factors into the area to determine the number of bags required.

Fill Depth	Fill Density		
	6 Lbs./C.F.	8 Lbs./C.F.	10 Lbs./C.F.
1″	85.0	63.8	51.0
2″	42.5	31.9	25.5
3″	28.4	21.3	17.0
3-1/2″	24.3	18.3	14.5
3-5/8″	23.5	17.6	14.1
4″	21.2	16.0	12.7
6″	14.2	10.6	8.5
10″	8.5	6.4	5.1

(courtesy, *Estimating Tables for Home Building*, Paul I. Thomas, Craftsman Book Company)

Batt Insulation Quantities

To determine the number of insulation batts needed, multiply the factor listed for the size batt to be used by the area (square feet) to be insulated. Note: S.F. area includes that occupied by studding or joists.

Batt Sizes	No.. of Batts per S.F.
15″ x 24″	.38
15″ x 48″	.19
19″ x 24″	.30
19″ x 48″	.15
23″ x 24″	.25
23″ x 48″	.125

(courtesy, *Estimating Tables for Home Building*, Paul I. Thomas, Craftsman Book Company)

Waste Values for Roof Styles

This table lists the approximate values for waste that should be added to the net area for materials of various roof types.

Roof Shape	Plain	Cut-up
Gable	10	15
Hip	15	20
Gambrel	10	20
Gothic	10	15
Mansard (sides)	10	15
Porches	10	—

Roofing Material Quantities — Shingles

This table describes various types of roofing shingles and some of their characteristics and expected coverages.

Product	Configuration	Approximate Shipping Weight per Square	Shingles per Square	Bundles per Square	Width	Length	Exposure	ASTM* Fire & Wind Ratings
Self-Sealing Random-Tab Strip Shingle Multi-Thickness	Various Edge, Surface Texture & Application Treatments	240# to 360#	64 to 90	3, 4 or 5	11-1/2" to 14"	36" to 40"	4" to 6"	A or C - Many Wind Resistant
Self-Sealing Random-Tab Strip Shingle Single Thickness	Various Edge, Surface Texture & Application Treatments	240# to 300#	65 to 80	3 or 4	12" to 13-1/4"	36" to 40"	4" to 5-5/8"	A or C - Many Wind Resistant
Self-Sealing Square-Tab Strip Shingle Three-Tab	3 Tab or 4 Tab	200# to 300#	65 to 80	3 or 4	12" to 13-1/4"	36" to 40"	5" to 5-5/8"	A or C - All Wind Resistant
Self-Sealing Square-Tab Strip Shingle No Cut-Out	Various Edge and Surface Texture Treatments	200# to 300#	65 to 81	3 or 4	12" to 13-1/4"	36" to 40"	5" to 5 5/8"	A or C - All Wind Resistant
Individual Interlocking Shingle Basic Design	Several Design Variations	180# to 250#	72 to 120	3 or 4	18" to 22-1/4"	20" to 22-1/2"	–	A or C - Many Wind Resistant

Other types available from some manufacturers in certain areas of the country. Consult your Regional Asphalt Roofing Manufacturers Association manufacturer.

* American Society for Testing and Materials.

(courtesy *Asphalt Roofing Manufacturers Association*)

Roofing Material Quantities — Red Cedar Shingles

This table shows expected coverages and materials required for the installation of various sizes of red cedar shingles.

	No. 1 Grade Sixteen Inch Shingles				No. 1 Grade Eighteen Inch Shingles				No. 1 Grade Twenty-Four Inch Shingles			
	Nail Size: 3d, 1-1/4" Long				Nail Size: 3d, 1-1/4" Long				Nail Size: 4d, 1-1/2" Long			
Shingles Exposure in Inches	Four-Bundle Square		One Bundle		Four Bundle: Square		One Bundle		Four-Bundle: Square		One Bundle	
	Coverage in S.F.	Pounds Nails	Coverage in S.F.	Pounds Nails	Coverage in S.F.	Pounds Nails	Coverage in S.F.	Pounds Nails	Coverage in S.F.	Pounds Nails	Coverage in S.F.	Pounds Nails
3-1/2	70	2-7/8	17-1/2	3/4								
4	80	2-1/2	20	5/8	72-1/2	2-1/2	18	5/8				
4-1/2	90	2-1/4	22-1/2	5/8	81-1/2	2-1/4	20	5/8				
5	100*	2	25	1/2	90-1/2	2	22-1/2	1/2				
5-1/2	110	1-3/4	27-1/2	1/2	100*	1-3/4	25	1/2				
6	120	1-2/3	30	3/8	109	1-2/3	27	3/8	80	2-1/3	20	5/8
6-1/2	130	1-1/2	32-1/2	3/8	118	1-1/2	29-1/2	3/8	86-1/2	2-1/8	21-1/2	1/2
7	140	1-2/5	35	1/3	127	1-2/5	31-1/2	1/3	93	2	23	1/2
7-1/2	150†	1-1/3	37-1/2	1/3	136	1-1/3	34	1/3	100*	1-7/8	25	1/2
8	160		40		145-1/2†	1-1/4	36	1/3	106-1/2	1-3/4	26-1/2	1/2
8-1/2	170		42-1/2		154-1/2	1-1/4	38-1/2	1/4	113	1-2/3	28	1/2
9	180		45		163-1/2		40-1/2		120	1-1/2	30	3/8
9-1/2	190		47-1/2		172-1/2		43		126-1/2	1-1/2	31-1/2	3/8
10	200		50		181-1/2		45		133	1-1/2	33	3/8
10-1/2	210		52-1/2		191		47-1/2		140	1-1/3	35	1/3
11	220		55		200		50		146-1/2	1-1/4	36-1/2	1/3
11-1/2	230		57-1/2		209		52		153†	1-1/4	38	1/3
12	240‡		60		218		54-1/2		160		40	
12-1/2					227		56-1/2		166-1/2		41-1/2	
13					236		59		173		43	
13-1/2					245-1/2		61		180		45	
14					254-1/2‡		63-1/2		186-1/2		46-1/2	
14-1/2									193		48	
15									200		50	
15-1/2									206-1/2		51-1/2	
16									213‡		53	

* Maximum exposure recommended for roofs.

† Maximum exposure recommended for single-coursing on side walls.

‡ Maximum exposure recommended for double-coursing on side walls. Figures in italics are inserted for convenience in estimating quantity of shingles needed for wide exposures in double-coursing, with butt-nailing. In double-coursing, with any exposure chosen, the figures indicate the amount of shingles for the outer courses. Order an equivalent number of shingles for concealed courses. Approximately 1-1/2 lbs. 5d small-headed nails required per square (100 S.F. wall area) to apply outer course of 16-inch shingles at 12-inch weather exposure. Plus 1/2 lb. 3d nails for under course shingles. Figure slightly fewer nails for 18-inch shingles at 14-inch exposure.

(from *Certigrade Handbook of Red Cedar Shingles*, published by Red Cedar Shingle Bureau (Red Cedar Shingle and Handsplit Shake Bureau), Seattle, Washington, 1957.)

(courtesy of *American Institute of Steel Construction, Inc.*)

Roofing Material Quantities — Standard 3/16″ Thick Slate

This table shows the materials needed for installing various sizes of 3/16″-thick roof slates.

Size of Slate (In.)	Slates per Square	Exposure with 3″ Lap	Nails per Square Lbs.	Nails per Square Ozs.	Size of Slate (In.)	Slates per Square	Exposure with 3″ Lap	Nails per Square Lbs.	Nails per Square Ozs.
26 x 14	89	11-1/2″	1	0	16 x 14	160	6-1/2″	1	13
					16 x 12	184	6-1/2″	2	2
24 x 16	86	10-1/2″	1	0	16 x 11	201	6-1/2″	2	5
24 x 14	98	10-1/2″	1	2	16 x 10	222	6-1/2″	2	8
24 x 13	106	10-1/2″	1	3	16 x 9	246	6-1/2″	2	13
24 x 12	114	10-1/2″	1	5	16 x 8	277	6-1/2″	3	2
24 x 11	125	10-1/2″	1	7					
					14 x 12	218	5-1/2″	2	8
22 x 14	108	9-1/2″	1	4	14 x 11	238	5-1/2″	2	11
22 x 13	117	9-1/2″	1	5	14 x 10	261	5-1/2″	3	3
22 x 12	126	9-1/2″	1	7	14 x 9	291	5-1/2″	3	5
22 x 11	138	9-1/2″	1	9	14 x 8	327	5-1/2″	3	12
22 x 10	152	9-1/2″	1	12	14 x 7	374	5-1/2″	4	4
20 x 14	121	8-1/2″	1	6	12 x 10	320	4-1/2″	3	10
20 x 13	132	8-1/2″	1	8	12 x 9	355	4-1/2″	4	1
20 x 12	141	8-1/2″	1	10	12 x 8	400	4-1/2″	4	9
20 x 11	154	8-1/2″	1	12	12 x 7	457	4-1/2″	5	3
20 x 10	170	8-1/2″	1	15	12 x 6	533	4-1/2″	6	1
20 x 9	189	8-1/2″	2	3					
					11 x 8	450	4″	5	2
18 x 14	137	7-1/2″	1	9	11 x 7	515	4″	5	14
18 x 13	148	7-1/2″	1	11					
18 x 12	160	7-1/2″	1	13	10 x 8	515	3-1/2″	5	14
18 x 11	175	7-1/2″	2	0	10 x 7	588	3-1/2″	7	4
18 x 10	192	7-1/2″	2	3	10 x 6	686	3-1/2″	7	
18 x 9	213	7-1/2″	2	7					

Roofing Material Quantities — Asphalt Rolls and Sheets

This table shows some of the characteristics of asphalt roofing materials.

Product	Approximate Shipping Weight per Roll	Approximate Shipping Weight per Square	Squares per Package	Length	Width	Selvage	Exposure	ASTM Fire & Wind Ratings
Mineral Surface Roll	75# to 90#	75# to 90#	1	36′ to 38′	36″	0″ to 4″	32″ to 34″	C
Mineral Surface Roll Double Coverage	55# to 70#	110# to 140#	1/2	36′	36″	19″	17″	C
Smooth Surface Roll	50# to 86#	40# to 65#	1 to 2	36′ to 72′	36″	N/A	34″	None
Saturated Felt Underlayment (non-perforated)	35# to 60#	11# to 30#	2 to 4	72′ to 144	36″	NA	17″ to 34″	*

*May be component in a complete fire-rated system. Check with manufacturer.
(courtesy *Asphalt Roofing Manufacturers Association*)

Wood Siding Quantities

This table allows you to compute the amount of wood siding (in board feet required per square foot of wall area to receive siding) based on the style and dimensions of the siding.

Type of Siding	Size	Exposure	Board Feet per S.F. of Wall Area	Lbs. Nails per MBM of Siding Stud Spacing 16"	20"	24"
Plain Bevel Siding	1/2" x 4"	2-1/2"	1.60	17	13	11
		2-3/4"	1.45			
	1/2" x 6"	4-1/2"	1.33	11	9	7
		4-3/4"	1.26			
		5"	1.20			
	1/2" x 8"	6-1/2"	1.23	8	7	6
		7"	1.14			
Plain Bevel Bungalow Siding	5/8" x 8"	6-1/2"	1.23	14	11	9
		7"	1.14			
	5/8" x 10"	8-1/2"	1.18	16	13	11
		9"	1.11			
	3/4" x 8"	6-1/2"	1.23	14	11	9
		7"	1.14			
	3/4" x 10"	8-1/2"	1.18	16	13	11
		9"	1.11			
	3/4" x 12"	10-1/2"	1.14	14	11	9
		11"	1.09			
Drop or Rustic Siding	3/4" x 4"	3-1/4"	1.23	27	22	18
	3/4" x 6"	5-1/6"	1.19	18	15	12
		5-3/16"	1.17			

Standard Door Nomenclature

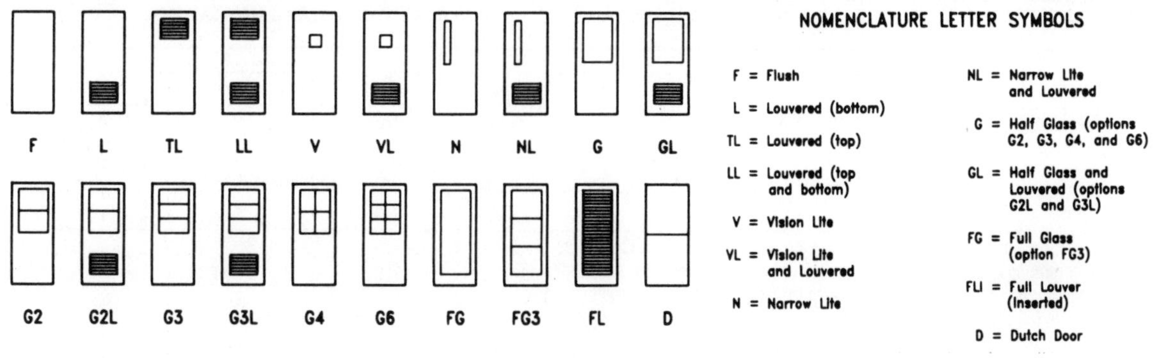

Weight of Doors

Door Thickness	Weight of Doors in Pounds per Square Foot				
	White Pine	Oak	Hollow Core	Solid Core	Hollow Metal
1-3/8"	3 psf	6 psf	1-1/2 psf	3-1/2 – 4 psf	6-1/2 psf
1-3/4"	3-1/2	7	2	4-1/2 – 5-1/4	6-1/2
2-1/4"	4-1/2	9	–	5-1/2 – 6-3/4	6-1/2

Finish Hardware Allowances

Average Allowances for Finish Hardware as a Percentage of Total Job Costs	
Minimum	0.75%
Maximum	3.50%

Average distribution of total finish hardware costs for a typical building is 85% material, 15% labor.

Fire Door Classifications

This table lists the various fire door ratings by label, time, and temperature rating, plus allowable glass area for a fire-rated door.

Classification	Time Rating (as shown on label)		Temperature Rise (as shown on label)	Maximum Glass Area
3 Hour fire doors (A) are for use in openings in walls separating buildings or dividing a single building into fire areas.	3 Hr.	(A)	30 Min. 250°F Max	None
	3 Hr.	(A)	30 Min. 450°F Max	
	3 Hr.	(A)	30 Min. 650°F Max	
	3 Hr.	(A)	*	
1-1/2 Hour Fire Doors (B) and (D) are for use in openings in 2 Hour enclosures of vertical communication through buildings (stairs, elevators, etc.) or in exterior walls which are subject to severe fire exposure from outside of the building. 1 Hour fire doors (B) are for use in openings in 1 Hour enclosures of vertical communication through buildings (stairs, elevators, etc.)	1-1/2 Hr.	(B)	30 Min. 250°F Max	100 square inches per door
	1-1/2 Hr.	(B)	30 Min. 450°F Max	
	1-1/2 Hr.	(B)	30 Min. 650°F Max	
	1-1/2 Hr.	(B)	*	
	1 Hr.	(B)	30 Min. 250°F Max	
	1-1/2 Hr.	(D)	30 Min. 250°F Max	None
	1-1/2 Hr.	(D)	30 Min. 450°F Max	
	1-1/2 Hr.	(D)	30 Min. 650°F Max	
	1-1/2 Hr.	(D)	*	
3/4 Hour fire doors (C) and (E) are for use in openings in corridor and room partitions or in exterior walls which are subject to moderate fire exposure from outside of the building.	3/4 Hr.	(C)	**	1296 square
	3/4 Hr.	(E)	**	720 square inches per light
1/2 Hour fire doors and 1/3 Hour fire doors are for use where smoke control is a primary consideration and are for the protection of openings in partitions between a habitable room and a corridor when the wall has a fire-resistance rating of not more than one hour.	1/2 Hr.		**	No Limit
	1/3		**	

* The labels do not record any temperature rise limits. This means that the temperature rise on the unexposed face of the door at the end of 30 minutes of test is in excess of 650°F.

** Temperature rise is not recorded

Quantities for Perlite Gypsum Plaster

This table shows the approximate number of sacks of prepared perlite gypsum plaster required to cover 100 square yards of surface area. The coverage depends on the type of lathing or base used and the required thickness of plaster.

Type of Base	Total Plaster Thickness (inches)	Number of Sacks
Wood lath	5/8	22 (80 lbs. per sack)
Metal lath	5/8	35 (80 lbs. per sack)
Gypsum lath	1/2	120 (80 lbs. per sack)
Masonry walls	5/8	24 (67 lbs. per sack)

(courtesy *How to Estimate Building Losses and Construction Costs*, Paul I. Thomas, Prentice-Hall, Inc.)

Quantities for White Coat Finish Plaster

This table shows the approximate quantities of materials required to cover 100 square yards of surface area with a 1/16" coat finish.

Type of Finish	100 Lb. Sacks Neat Gypsum	100 Lb. Sacks Keene's Cement	50 Lb. Sacks Hydrated Lime
Lime putty	2	–	8
Keene's cement	–	4	4

Based on ASA mix of 1 (100-lb.) sack neat gypsum gauging plaster to 4 (50-lb) sacks hydrated lime, and 1 (100-lb.) sack of Keene's cement to 1 (50-lb.) sack hydrated lime for medium hard finish.

(courtesy *How to Estimate Building Losses and Construction Costs*, Paul I. Thomas, Prentice-Hall, Inc.)

Quantities for Portland Cement Plaster

This table shows the approximate quantities of materials required to cover 100 square yards of surface area with Portland cement plaster. The coverage depends on the base material used and the required thickness of plaster.

Plaster Thickness (inches)	Cubic Yards on Masonry Base	Cubic Yards on Wire Lath Base
1/4	0.75	0.90
1/2	1.50	1.80
5/8	1.37	1.64
3/4	2.25	2.70
1	3.00	3.60

(courtesy *How to Estimate Building Losses and Construction Costs*, Paul I. Thomas, Prentice Hall, Inc.)

Quantities for Job-Mixed Plaster

This table presents approximate quantities required to prepare enough job-mixed sanded or perlite plaster to cover 100 square yards of surface area. The coverage depends on the type of base material used and the required thickness of plaster.

Type of Base	Total Plaster Thickness Inches	100 Lb. Sacks of Neat Gypsum	C.Y. Sand 1:2-1/2 Mix	Aggregate Perlite		
				C.F.	4 C.F. Sacks	Mix
Wood lath	5/8	11	1.1	27.5	12.5	1:2-1/2
Metal lath	5/8	20	2.0	50.0	12.5	1:2-1/2
Gypsum lath	1/2	10	1.0	25	6.5	1:2-1/2
Masonry walls	5/8	12	1.2	30	7.5	1:3

ASA permits 250 lbs. damp loose sand or 2-1/2 C.F. of vermiculite or perlite, provided this proportioning is used for both scratch and brown coats on three-coat work.
The number of 4 C.F. sacks of perlite are shown to nearest 1/2 sack.

(courtesy *How to Estimate Building Losses and Construction Costs*, Paul I. Thomas, Prentice-Hall, Inc.

Furring Quantities

This table provides multiplication factors for converting square feet of wall area requiring furring into board feet of lumber. These figures are based on lumber size and spacing.

| | Board Feet per Square Feet of Wall Area | | | | |
| | Spacing Center to Center | | | | Lbs. Nails per MBM of Furring |
Size	12"	16"	20"	24"	
1" x 2"	.18	.14	.11	.10	55
1" x 3"	.28	.21	.17	.14	37

Gypsum Panel Coverage

This table shows the expected coverage from the size panels indicated across the top of the table. Inversely, if the area to be covered with gypsum board panels is known (in square feet), you can quickly determine the required number of sheets of drywall. Note that no provision has been made for openings.

| | Sizes and Numbers of Panels and S.F. of Area Covered | | | | | | | |
No. of Panels	4' x 7'	4' x 8'	4' x 9'	4' x 10'	4' x 11'	4' x 12'	4' x 13'	4' x 14'
10	280	320	360	400	440	480	520	560
11	308	352	396	440	484	528	572	616
12	336	384	432	480	528	576	624	672
13	364	416	468	520	572	624	676	728
14	392	448	504	560	616	672	728	784
15	420	480	540	600	660	720	780	840
16	448	512	576	640	704	768	832	896
17	476	544	612	680	748	816	884	952
18	504	576	648	720	792	864	936	1008
19	532	608	684	760	836	912	988	1064
20	560	640	720	800	880	960	1040	1120
21	588	672	756	840	924	1008	1092	1176
22	616	704	792	880	968	1056	1144	1232
23	644	736	828	920	1012	1104	1196	1288
24	672	768	864	960	1056	1152	1248	1344
25	700	800	900	1000	1100	1200	1300	1400
26	728	832	936	1040	1144	1248	1352	1456
27	756	864	972	1080	1188	1296	1404	1512
28	784	896	1008	1120	1232	1344	1456	1568
29	812	928	1044	1160	1276	1392	1508	1624
30	840	960	1080	1200	1320	1440	1560	1680
31	868	992	1116	1240	1364	1488	1612	1736

Drywall Accessories

This table provides the approximate quantities of nails, joint compound and tape needed for the indicated amounts of drywall.

With this Amount of Sheetrock Gypsum Panel	Type GWB-54 Nails Required*	USE	this Amount of Powder-Type Compound	OR	this Amount of USG Ready-to-Use Compound All Purpose	AND	this Amount of Perf-A-Tape Reinforcing Tape
100 S.F.	.6 lbs.		6 lbs.		1 Gal.		37 Ft.
200 S.F.	1.1 lbs.		12 lbs.		2 Gals.		74 Ft.
300 S.F.	1.6 lbs.		18 lbs.		2 Gals.		111 Ft.
400 S.F.	2.1 lbs.		24 lbs.		3 Gals.		148 Ft.
500 S.F.	2.7 lbs.		30 lbs.		3 Gals.		185 Ft.
600 S.F.	3.2 lbs.		36 lbs.		4 Gals.		222 Ft.
700 S.F.	3.7 lbs.		42 lbs.		5 Gals.		259 Ft.
800 S.F.	4.2 lbs.		48 lbs.		5 Gals.		296 Ft.
900 S.F.	4.8 lbs.		54 lbs.		6 Gals.		333 Ft.
1000 S.F.	5.3 lbs.		60 lbs.		6 Gals.		370 Ft.

* Spaced 7″ on ceiling; 8″ on wall.

Wallboard Quantities

This table lists factors that can be applied to predetermined area figures in order to quantify actual material requirements. Included are waste factors for the intended use of the materials. The table also lists the amount of nails required for the wallboard, based on stud spacing.

Includes Fiber Board, Gypsum Board, and Plywood, Used as Underflooring, Sheathing, Plaster Base, or as Drywall Finish					
Factors		Nails			
Used for Underflooring, Sheathing, and Plaster Base	Used for Exposed Drywall Finish	Pounds of Nails per 1000 Sq. Ft. of Wallboard			
		Joist, Stud or Rafter Spacing			
		12″	16″	20″	24″
1.05	1.10	7	6	5	4

Location Factors

Costs shown in *Contractor's Pricing Guide: Framing & Rough Carpentry 1996* are based on National Averages for materials and installation. To adjust these costs to a specific location, simply multiply the base cost by the factor for that city. The data is arranged alphabetically by state and postal zip code numbers. For a city not listed, use the factor for a nearby city with similar economic characteristics.

STATE/ZIP	CITY	Residential	Commercial
ALABAMA			
350-352	Birmingham	.83	.84
354	Tuscaloosa	.81	.79
355	Jasper	.76	.77
356	Decatur	.83	.84
357-358	Huntsville	.82	.83
359	Gadsden	.81	.82
360-361	Montgomery	.82	.80
362	Anniston	.74	.75
363	Dothan	.82	.80
364	Evergreen	.82	.80
365-366	Mobile	.84	.85
367	Selma	.82	.80
368	Phenix City	.82	.80
369	Butler	.82	.80
ALASKA			
995-996	Anchorage	1.30	1.29
997	Fairbanks	1.29	1.28
998	Juneau	1.29	1.28
999	Ketchikan	1.34	1.33
ARIZONA			
850,853	Phoenix	.94	.91
852	Mesa/Tempe	.89	.86
855	Globe	.93	.90
856-857	Tucson	.93	.90
859	Show Low	.94	.90
860	Flagstaff	.96	.92
863	Prescott	.93	.89
864	Kingman	.93	.89
865	Chambers	.93	.89
ARKANSAS			
716	Pine Bluff	.80	.80
717	Camden	.72	.72
718	Texarkana	.77	.76
719	Hot Springs	.72	.72
720-722	Little Rock	.80	.80
723	West Memphis	.82	.82
724	Jonesboro	.82	.82
725	Batesville	.78	.78
726	Harrison	.79	.79
727	Fayetteville	.71	.68
728	Russellville	.80	.77
729	Fort Smith	.83	.80
749	Poteau	.84	.80
CALIFORNIA			
900-902	Los Angeles	1.13	1.13
903-905	Inglewood	1.11	1.11
906-908	Long Beach	1.12	1.12
910-912	Pasadena	1.11	1.11
913-916	Van Nuys	1.13	1.13
917-918	Alhambra	1.12	1.12
919-921	San Diego	1.13	1.09
922	Palm Springs	1.15	1.11
923-924	San Bernardino	1.13	1.09
925	Riverside	1.16	1.12
926-927	Santa Ana	1.14	1.11
928	Anaheim	1.14	1.12
930	Oxnard	1.18	1.12
931	Santa Barbara	1.14	1.11
932-933	Bakersfield	1.14	1.08
934	San Luis Obispo	1.25	1.12
935	Mojave	1.13	1.09
936-938	Fresno	1.15	1.11
939	Salinas	1.14	1.14
940-941	San Francisco	1.23	1.26
942,956-958	Sacramento	1.14	1.13
943	Palo Alto	1.16	1.19
944	San Mateo	1.17	1.19
945	Vallejo	1.15	1.18
946	Oakland	1.17	1.20
947	Berkeley	1.30	1.34
948	Richmond	1.16	1.19
949	San Rafael	1.27	1.21
950	Santa Cruz	1.19	1.17
951	San Jose	1.24	1.22
952	Stockton	1.16	1.12
953	Modesto	1.16	1.12
954	Santa Rosa	1.17	1.21
955	Eureka	1.14	1.13
959	Marysville	1.14	1.13
960	Redding	1.12	1.11
961	Susanville	1.12	1.11
COLORADO			
800-802	Denver	.97	.93
803	Boulder	.91	.87
804	Golden	.95	.91

STATE/ZIP	CITY	Residential	Commercial
COLORADO (CONT'D)			
805	Fort Collins	.99	.92
806	Greeley	.92	.86
807	Fort Morgan	.97	.91
808-809	Colorado Springs	.92	.90
810	Pueblo	.93	.91
811	Alamosa	.91	.89
812	Salida	.91	.89
813	Durango	.89	.87
814	Montrose	.87	.85
815	Grand Junction	.92	.87
816	Glenwood Springs	.97	.92
CONNECTICUT			
060	New Britain	1.07	1.08
061	Hartford	1.08	1.09
062	Willimantic	1.08	1.09
063	New London	1.08	1.07
064	Meriden	1.07	1.08
065	New Haven	1.07	1.08
066	Bridgeport	1.05	1.08
067	Waterbury	1.08	1.08
068	Norwalk	1.04	1.08
069	Stamford	1.06	1.10
D.C.			
200-205	Washington	.95	.97
DELAWARE			
197	Newark	.99	1.00
198	Wilmington	.99	1.00
199	Dover	.99	1.00
FLORIDA			
320,322	Jacksonville	.86	.85
321	Daytona Beach	.90	.89
323	Tallahassee	.79	.81
324	Panama City	.73	.74
325	Pensacola	.88	.86
326	Gainesville	.87	.84
327-328,347	Orlando	.90	.88
329	Melbourne	.91	.90
330-332,340	Miami	.85	.87
333	Fort Lauderdale	.86	.88
334,349	West Palm Beach	.88	.85
335-336,346	Tampa	.84	.86
337	St. Petersburg	.85	.87
338	Lakeland	.83	.85
339	Fort Myers	.83	.83
342	Sarasota	.83	.85
GEORGIA			
300-303,399	Atlanta	.81	.86
304	Statesboro	.66	.68
305	Gainesville	.62	.66
306	Athens	.72	.76
307	Dalton	.67	.67
308-309	Augusta	.78	.80
310-312	Macon	.81	.81
313-314	Savannah	.80	.81
315	Waycross	.75	.75
316	Valdosta	.77	.77
317	Albany	.78	.80
318-319	Columbus	.77	.77
HAWAII			
967	Hilo	1.26	1.22
968	Honolulu	1.27	1.23
STATES & POSS.			
969	Guam	.87	.84
IDAHO			
832	Pocatello	.95	.94
833	Twin Falls	.82	.81
834	Idaho Falls	.86	.85
835	Lewiston	1.11	1.02
836-837	Boise	.94	.93
838	Coeur d'Alene	1.00	.93
ILLINOIS			
600-603	North Suburban	1.08	1.07
604	Joliet	1.07	1.06
605	South Suburban	1.08	1.07
606	Chicago	1.09	1.08
609	Kankakee	1.00	1.00
610-611	Rockford	1.03	1.02
612	Rock Island	1.04	.96
613	La Salle	1.07	1.00
614	Galesburg	1.04	.97
615-616	Peoria	1.07	1.00

STATE/ZIP	CITY	Residential	Commercial
ILLINOIS (CONT'D)			
617	Bloomington	1.01	.97
618-619	Champaign	1.02	.99
620-622	East St. Louis	1.00	1.00
623	Quincy	.96	.94
624	Effingham	.97	.94
625	Decatur	1.01	.98
626-627	Springfield	1.01	.98
628	Centralia	.98	.98
629	Carbondale	.96	.96
INDIANA			
424	Henderson	.94	.92
460	Anderson	.93	.92
461-462	Indianapolis	.97	.95
463-464	Gary	1.01	.99
465-466	South Bend	.92	.90
467-468	Fort Wayne	.91	.92
469	Kokomo	.90	.89
470	Lawrenceburg	.90	.88
471	New Albany	.90	.86
472	Columbus	.90	.88
473	Muncie	.92	.91
474	Bloomington	.92	.90
475	Washington	.90	.90
476-477	Evansville	.94	.94
478	Terre Haute	.93	.92
479	Lafayette	.90	.90
IOWA			
500-503,509	Des Moines	.94	.90
504	Mason City	.88	.82
505	Fort Dodge	.85	.79
506-507	Waterloo	.89	.83
508	Creston	.91	.87
510-511	Sioux City	.89	.83
512	Sibley	.82	.80
513	Spencer	.83	.81
514	Carroll	.88	.83
515	Council Bluffs	.93	.87
516	Shenandoah	.76	.72
520	Dubuque	.96	.86
521	Decorah	.92	.83
522-524	Cedar Rapids	.99	.90
525	Ottumwa	.94	.86
526	Burlington	.84	.79
527-528	Davenport	.93	.91
KANSAS			
660-662	Kansas City	.96	.94
664-666	Topeka	.87	.86
667	Fort Scott	.87	.85
668	Emporia	.78	.77
669	Belleville	.91	.85
670-672	Wichita	.89	.86
673	Independence	.82	.79
674	Salina	.87	.83
675	Hutchinson	.80	.77
676	Hays	.89	.85
677	Colby	.89	.85
678	Dodge City	.89	.86
679	Liberal	.81	.79
KENTUCKY			
400-402	Louisville	.93	.90
403-405	Lexington	.90	.86
406	Frankfort	.96	.90
407-409	Corbin	.81	.76
410	Covington	.96	.93
411-412	Ashland	.94	.95
413-414	Campton	.80	.76
415-416	Pikeville	.85	.85
417-418	Hazard	.79	.76
420	Paducah	.96	.91
421-422	Bowling Green	.94	.89
423	Owensboro	.92	.90
425-426	Somerset	.78	.75
427	Elizabethtown	.92	.88
LOUISIANA			
700-701	New Orleans	.87	.86
703	Thibodaux	.86	.86
704	Hammond	.85	.84
705	Lafayette	.87	.84
706	Lake Charles	.87	.87
707-708	Baton Rouge	.85	.84
710-711	Shreveport	.82	.82
712	Monroe	.80	.80
713-714	Alexandria	.80	.80
MAINE			
039	Kittery	.81	.83
040-041	Portland	.89	.92
042	Lewiston	.91	.92
043	Augusta	.82	.82
044	Bangor	.94	.94
045	Bath	.81	.81

STATE/ZIP	CITY	Residential	Commercial
MAINE (CONT'D)			
046	Machias	.78	.78
047	Houlton	.83	.83
048	Rockland	.86	.86
049	Waterville	.82	.81
MARYLAND			
206	Waldorf	.88	.88
207-208	College Park	.90	.90
209	Silver Spring	.89	.89
210-212	Baltimore	.91	.91
214	Annapolis	.89	.90
215	Cumberland	.87	.88
216	Easton	.70	.71
217	Hagerstown	.90	.89
218	Salisbury	.78	.78
219	Elkton	.82	.83
MASSACHUSETTS			
010-011	Springfield	1.08	1.06
012	Pittsfield	1.04	1.04
013	Greenfield	1.05	1.03
014	Fitchburg	1.13	1.09
015-016	Worcester	1.15	1.10
017	Framingham	1.11	1.12
018	Lowell	1.13	1.13
019	Lawrence	1.12	1.12
020-022	Boston	1.18	1.19
023-024	Brockton	1.10	1.12
025	Buzzards Bay	1.07	1.09
026	Hyannis	1.09	1.10
027	New Bedford	1.11	1.12
MICHIGAN			
480,483	Royal Oak	1.03	1.02
481	Ann Arbor	1.04	1.04
482	Detroit	1.06	1.05
484-485	Flint	1.00	1.01
486	Saginaw	.97	.98
487	Bay City	.95	.96
488-489	Lansing	1.02	.99
490	Battle Creek	1.02	.96
491	Kalamazoo	1.01	.95
492	Jackson	.97	.94
493,495	Grand Rapids	.90	.87
494	Muskegan	.97	.94
496	Traverse City	.94	.91
497	Gaylord	.94	.95
498-499	Iron Mountain	.98	.95
MINNESOTA			
540	New Richmond	.99	.92
550-551	Saint Paul	1.08	1.05
553-554	Minneapolis	1.14	1.11
556-558	Duluth	.98	.99
559	Rochester	1.04	1.01
560	Mankato	.95	.94
561	Windom	.80	.79
562	Willmar	.84	.83
563	St. Cloud	1.07	.99
564	Brainerd	1.02	.95
565	Detroit Lakes	.86	.93
566	Bemidji	.86	.92
567	Thief River Falls	.83	.89
MISSISSIPPI			
386	Clarksdale	.71	.67
387	Greenville	.82	.78
388	Tupelo	.72	.73
389	Greenwood	.73	.69
390-392	Jackson	.83	.79
393	Meridian	.77	.76
394	Laurel	.73	.70
395	Biloxi	.84	.80
396	Mc Comb	.69	.68
397	Columbus	.72	.73
MISSOURI			
630-631	St. Louis	.98	1.01
633	Bowling Green	.93	.96
634	Hannibal	1.02	.95
635	Kirksville	.86	.90
636	Flat River	.94	.97
637	Cape Girardeau	.95	.97
638	Sikeston	.84	.86
639	Poplar Bluff	.89	.91
640-641	Kansas City	.98	.95
644-645	St. Joseph	.87	.91
646	Chillicothe	.80	.84
647	Harrisonville	.94	.92
648	Joplin	.85	.87
650-651	Jefferson City	.97	.90
652	Columbia	.95	.89
653	Sedalia	.94	.87
654-655	Rolla	.88	.82
656-658	Springfield	.84	.86

Location Factors

STATE/ZIP	CITY	Residential	Commercial
MONTANA			
590-591	Billings	1.01	.98
592	Wolf Point	.99	.97
593	Miles City	1.00	.98
594	Great Falls	1.00	.99
595	Havre	.97	.96
596	Helena	.99	.98
597	Butte	.98	.97
598	Missoula	.98	.97
599	Kalispell	.97	.96
NEBRASKA			
680-681	Omaha	.90	.89
683-685	Lincoln	.87	.82
686	Columbus	.77	.76
687	Norfolk	.87	.86
688	Grand Island	.86	.82
689	Hastings	.86	.82
690	Mc Cook	.76	.72
691	North Platte	.86	.82
692	Valentine	.80	.76
693	Alliance	.77	.73
NEVADA			
889-891	Las Vegas	1.04	1.03
893	Ely	.97	.98
894-895	Reno	.95	1.00
897	Carson City	.97	1.00
898	Elko	.94	.97
NEW HAMPSHIRE			
030	Nashua	.97	.98
031	Manchester	.97	.98
032-033	Concord	.96	.97
034	Keene	.84	.85
035	Littleton	.85	.86
036	Charleston	.82	.82
037	Claremont	.81	.82
038	Portsmouth	.96	.95
NEW JERSEY			
070-071	Newark	1.14	1.12
072	Elizabeth	1.12	1.10
073	Jersey City	1.14	1.13
074-075	Paterson	1.14	1.14
076	Hackensack	1.12	1.12
077	Long Branch	1.12	1.10
078	Dover	1.14	1.12
079	Summit	1.12	1.10
080,083	Vineland	1.12	1.08
081	Camden	1.12	1.09
082,084	Atlantic City	1.12	1.09
085-086	Trenton	1.14	1.12
087	Point Pleasant	1.12	1.10
088-089	New Brunswick	1.14	1.12
NEW MEXICO			
870-872	Albuquerque	.88	.90
873	Gallup	.89	.91
874	Farmington	.89	.91
875	Santa Fe	.88	.90
877	Las Vegas	.88	.90
878	Socorro	.89	.91
879	Truth/Consequences	.88	.88
880	Las Cruces	.85	.85
881	Clovis	.90	.90
882	Roswell	.91	.91
883	Carrizozo	.91	.91
884	Tucumcari	.91	.91
NEW YORK			
100-102	New York	1.36	1.36
103	Staten Island	1.31	1.31
104	Bronx	1.31	1.31
105	Mount Vernon	1.24	1.24
106	White Plains	1.23	1.23
107	Yonkers	1.26	1.26
108	New Rochelle	1.25	1.25
109	Suffern	1.17	1.17
110	Queens	1.30	1.30
111	Long Island City	1.31	1.31
112	Brooklyn	1.31	1.31
113	Flushing	1.31	1.31
114	Jamaica	1.30	1.30
115,117,118	Hicksville	1.28	1.28
116	Far Rockaway	1.25	1.25
119	Riverhead	1.28	1.28
120-122	Albany	1.00	1.00
123	Schenectady	1.01	1.01
124	Kingston	1.15	1.13
125-126	Poughkeepsie	1.18	1.16
127	Monticello	1.14	1.12
128	Glens Falls	.97	.95
130-132	Syracuse	1.01	.99
133-135	Utica	.93	.96
136	Watertown	.96	.99
137-139	Binghamton	.97	.97

STATE/ZIP	CITY	Residential	Commercial
NEW YORK (CONT'D)			
140-142	Buffalo	1.09	1.05
143	Niagara Falls	1.09	1.05
144-146	Rochester	1.03	1.04
147	Jamestown	.94	.91
148-149	Elmira	.98	.96
NORTH CAROLINA			
270,272-274	Greensboro	.78	.79
271	Winston-Salem	.78	.79
275-276	Raleigh	.79	.79
277	Durham	.78	.79
278	Rocky Mount	.65	.65
279	Elizabeth City	.67	.67
280	Gastonia	.77	.78
281-282	Charlotte	.77	.78
283	Fayetteville	.79	.79
284	Wilmington	.76	.78
285	Kinston	.67	.67
286	Hickory	.65	.65
287-288	Asheville	.76	.78
289	Murphy	.66	.67
NORTH DAKOTA			
580-581	Fargo	.79	.84
582	Grand Forks	.80	.85
583	Devils Lake	.80	.85
584	Jamestown	.80	.85
585	Bismarck	.81	.85
586	Dickinson	.81	.85
587	Minot	.81	.85
588	Williston	.80	.84
OHIO			
430-432	Columbus	.95	.93
433	Marion	.91	.92
434-436	Toledo	.98	.97
437-438	Zanesville	.91	.90
439	Steubenville	.96	.96
440	Lorain	1.02	.96
441	Cleveland	1.08	1.02
442-443	Akron	1.00	.99
444-445	Youngstown	.99	.96
446-447	Canton	.95	.94
448-449	Mansfield	.95	.93
450	Hamilton	.98	.92
451-452	Cincinnati	.98	.92
453-454	Dayton	.92	.91
455	Springfield	.93	.91
456	Chillicothe	1.00	.94
457	Athens	.90	.89
458	Lima	.93	.92
OKLAHOMA			
730-731	Oklahoma City	.82	.84
734	Ardmore	.82	.81
735	Lawton	.83	.82
736	Clinton	.80	.81
737	Enid	.83	.82
738	Woodward	.82	.81
739	Guymon	.71	.70
740-741	Tulsa	.87	.84
743	Miami	.85	.82
744	Muskogee	.78	.76
745	Mc Alester	.76	.78
746	Ponca City	.81	.80
747	Durant	.79	.81
748	Shawnee	.78	.80
OREGON			
970-972	Portland	1.08	1.06
973	Salem	1.06	1.05
974	Eugene	1.05	1.04
975	Medford	1.06	1.05
976	Klamath Falls	1.06	1.05
977	Bend	1.05	1.04
978	Pendleton	1.04	1.01
979	Vale	.99	.97
PENNSYLVANIA			
150-152	Pittsburgh	1.04	1.02
153	Washington	1.02	1.00
154	Uniontown	1.01	.99
155	Bedford	1.04	.97
156	Greensburg	1.02	1.00
157	Indiana	1.05	.98
158	Dubois	1.04	.98
159	Johnstown	1.05	.98
160	Butler	1.01	.98
161	New Castle	1.01	.98
162	Kittanning	1.03	1.00
163	Oil City	.91	.95
164-165	Erie	.99	.98
166	Altoona	1.06	.97
167	Bradford	1.00	.98
168	State College	.98	.98
169	Wellsboro	.93	.94

STATE/ZIP	CITY	Residential	Commercial
PENNSYLVANIA (CONT'D)			
170-171	Harrisburg	.98	.97
172	Chambersburg	.97	.96
173-174	York	.99	.97
175-176	Lancaster	.98	.96
177	Williamsport	.94	.93
178	Sunbury	.96	.95
179	Pottsville	.96	.95
180	Lehigh Valley	1.03	1.02
181	Allentown	1.03	1.02
182	Hazleton	.97	.97
183	Stroudsburg	.99	.98
184-185	Scranton	.97	1.00
186-187	Wilkes-Barre	.94	.97
188	Montrose	.93	.96
189	Doylestown	.93	1.05
190-191	Philadelphia	1.12	1.10
193	Westchester	1.05	1.04
194	Norristown	1.07	1.05
195-196	Reading	.98	.99
RHODE ISLAND			
028	Newport	1.06	1.08
029	Providence	1.06	1.08
SOUTH CAROLINA			
290-292	Columbia	.75	.78
293	Spartanburg	.74	.77
294	Charleston	.76	.78
295	Florence	.74	.76
296	Greenville	.74	.77
297	Rock Hill	.65	.68
298	Aiken	.66	.68
299	Beaufort	.69	.71
SOUTH DAKOTA			
570-571	Sioux Falls	.89	.83
572	Watertown	.87	.81
573	Mitchell	.87	.81
574	Aberdeen	.88	.82
575	Pierre	.87	.81
576	Mobridge	.88	.81
577	Rapid City	.86	.80
TENNESSEE			
370-372	Nashville	.84	.84
373-374	Chattanooga	.85	.84
375,380-381	Memphis	.86	.86
376	Johnson City	.83	.82
377-379	Knoxville	.82	.82
382	Mc Kenzie	.71	.71
383	Jackson	.70	.77
384	Columbia	.78	.78
385	Cookeville	.70	.70
TEXAS			
750	Mc Kinney	.91	.84
751	Waxahachie	.84	.84
752-753	Dallas	.90	.86
754	Greenville	.81	.75
755	Texarkana	.91	.80
756	Longview	.86	.75
757	Tyler	.92	.81
758	Palestine	.75	.75
759	Lufkin	.80	.80
760-761	Fort Worth	.85	.84
762	Denton	.90	.82
763	Wichita Falls	.83	.83
764	Eastland	.77	.76
765	Temple	.80	.79
766-767	Waco	.83	.82
768	Brownwood	.76	.75
769	San Angelo	.82	.78
770-772	Houston	.89	.90
773	Huntsville	.84	.84
774	Wharton	.80	.81
775	Galveston	.88	.89
776-777	Beaumont	.87	.88
778	Bryan	.84	.85
779	Victoria	.84	.84
780	Laredo	.80	.81
781-782	San Antonio	.82	.83
783-784	Corpus Christi	.83	.82
785	Mc Allen	.82	.80
786-787	Austin	.81	.84
788	Del Rio	.71	.71
789	Giddings	.76	.76
790-791	Amarillo	.82	.82
792	Childress	.78	.81
793-794	Lubbock	.81	.83
795-796	Abilene	.80	.80
797	Midland	.82	.83
798-799,885	El Paso	.81	.80
UTAH			
840-841	Salt Lake City	.88	.87
842,844	Ogden	.89	.87

STATE/ZIP	CITY	Residential	Commercial
UTAH (CONT'D)			
843	Logan	.90	.88
845	Price	.82	.82
846-847	Provo	.89	.88
VERMONT			
050	White River Jct.	.76	.75
051	Bellows Falls	.77	.76
052	Bennington	.73	.72
053	Brattleboro	.77	.77
054	Burlington	.85	.86
056	Montpelier	.84	.85
057	Rutland	.87	.86
058	St. Johnsbury	.78	.79
059	Guildhall	.77	.78
129	Plattsburgh	.98	.96
VIRGINIA			
220-221	Fairfax	.89	.90
222	Arlington	.90	.91
223	Alexandria	.91	.92
224-225	Fredericksburg	.87	.88
226	Winchester	.81	.82
227	Culpeper	.80	.81
228	Harrisonburg	.77	.77
229	Charlottesville	.85	.83
230-232	Richmond	.86	.84
233-235	Norfolk	.83	.83
236	Newport News	.84	.83
237	Portsmouth	.82	.82
238	Petersburg	.86	.84
239	Farmville	.78	.76
240-241	Roanoke	.80	.79
242	Bristol	.81	.77
243	Pulaski	.73	.72
244	Staunton	.75	.73
245	Lynchburg	.83	.79
246	Grundy	.72	.72
WASHINGTON			
980-981,987	Seattle	1.01	1.06
982	Everett	.99	1.05
983-984	Tacoma	1.07	1.05
985	Olympia	1.07	1.05
986	Vancouver	1.11	1.04
988	Wenatchee	.98	1.02
989	Yakima	1.05	1.03
990-992	Spokane	1.01	1.00
993	Richland	1.02	1.01
994	Clarkston	1.01	1.00
WEST VIRGINIA			
247-248	Bluefield	.86	.86
249	Lewisburg	.83	.83
250-253	Charleston	.92	.92
254	Martinsburg	.77	.77
255-257	Huntington	.92	.94
258-259	Beckley	.92	.92
260	Wheeling	.91	.93
261	Parkersburg	.89	.91
262	Buckhannon	.94	.91
263-264	Clarksburg	.94	.92
265	Morgantown	.94	.92
266	Gassaway	.92	.92
267	Romney	.90	.90
268	Petersburg	.93	.91
WISCONSIN			
530,532	Milwaukee	1.00	.99
531	Kenosha	.96	.95
534	Racine	1.02	.97
535	Beloit	.99	.97
537	Madison	.96	.94
538	Lancaster	.90	.88
539	Portage	.88	.86
541-543	Green Bay	.98	.95
544	Wausau	.96	.92
545	Rhinelander	.98	.94
546	La Crosse	.95	.92
547	Eau Claire	1.01	.93
548	Superior	1.01	.95
549	Oshkosh	.95	.92
WYOMING			
820	Cheyenne	.89	.84
821	Yellow. Nat'l Park	.84	.81
822	Wheatland	.84	.79
823	Rawlins	.85	.80
824	Worland	.82	.79
825	Riverton	.83	.80
826	Casper	.88	.84
827	Newcastle	.83	.79
828	Sheridan	.87	.84
829-831	Rock Springs	.86	.81

STATE/ZIP	CITY	Residential	Commercial
CANADIAN FACTORS (reflect Canadian currency)			
ALBERTA			
	Calgary	1.03	1.00
	Edmonton	1.03	1.00
BRITISH COLUMBIA			
	Vancouver	1.09	1.10
	Victoria	1.06	1.07
MANITOBA			
	Winnipeg	1.02	1.01
NEW BRUNSWICK			
	Moncton	.96	.94
	Saint John	1.00	.98
NEWFOUNDLAND			
	St. John's	.98	.97
NOVA SCOTIA			
	Halifax	1.00	.99
ONTARIO			
	Barrie	1.14	1.12
	Hamilton	1.17	1.13
	Kitchener	1.09	1.07
	London	1.13	1.11
	Oshawa	1.14	1.12
	Ottawa	1.13	1.11
	Owen Sound	1.13	1.11
	St. Catherines	1.09	1.07
	Sudbury	1.08	1.06
	Thunder Bay	1.09	1.07
	Toronto	1.15	1.14
	Windsor	1.10	1.08
PRINCE EDWARD ISLAND			
	Charlottetown	.95	.93
QUEBEC			
	Chicoutimi	1.04	1.03
	Montreal	1.11	1.04
	Quebec	1.13	1.05
SASKATCHEWAN			
	Regina	.93	.93
	Saskatoon	.93	.93

Abbreviations

A	Area
ASTM	American Society for Testing and Materials
B.F.	Board feet
Carp.	Carpenter
C.F.	Cubic feet
CWJ	Composite wood joist
C.Y.	Cubic yard
Ea.	Each
Equip.	Equipment
Exp.	Exposure
Ext.	Exterior
F	Fahrenheit
Ft.	Foot, feet
Gal.	Gallon
Hr.	Hour
in.	Inch, inches
Inst.	Installation
Int.	Interior
Lb.	Pound
L.F.	Linear feet
LVL	Laminated veneer lumber
Mat.	Material
Max.	Maximum
MBF	Thousand board feet
MBM	Thousand feet board measure
MSF	Thousand square feet
Min.	Minimum
O.C.	On center
O&P	Overhead and profit
OWJ	Open web wood joist
Oz.	Ounce
Pr.	Pair
Quan.	Quantity
S.F.	Square foot
Sq.	Square, 100 square feet
S.Y.	Square yard
V.L.F.	Vertical linear feet
'	Foot, feet
"	Inch, inches
°	Degrees

Index

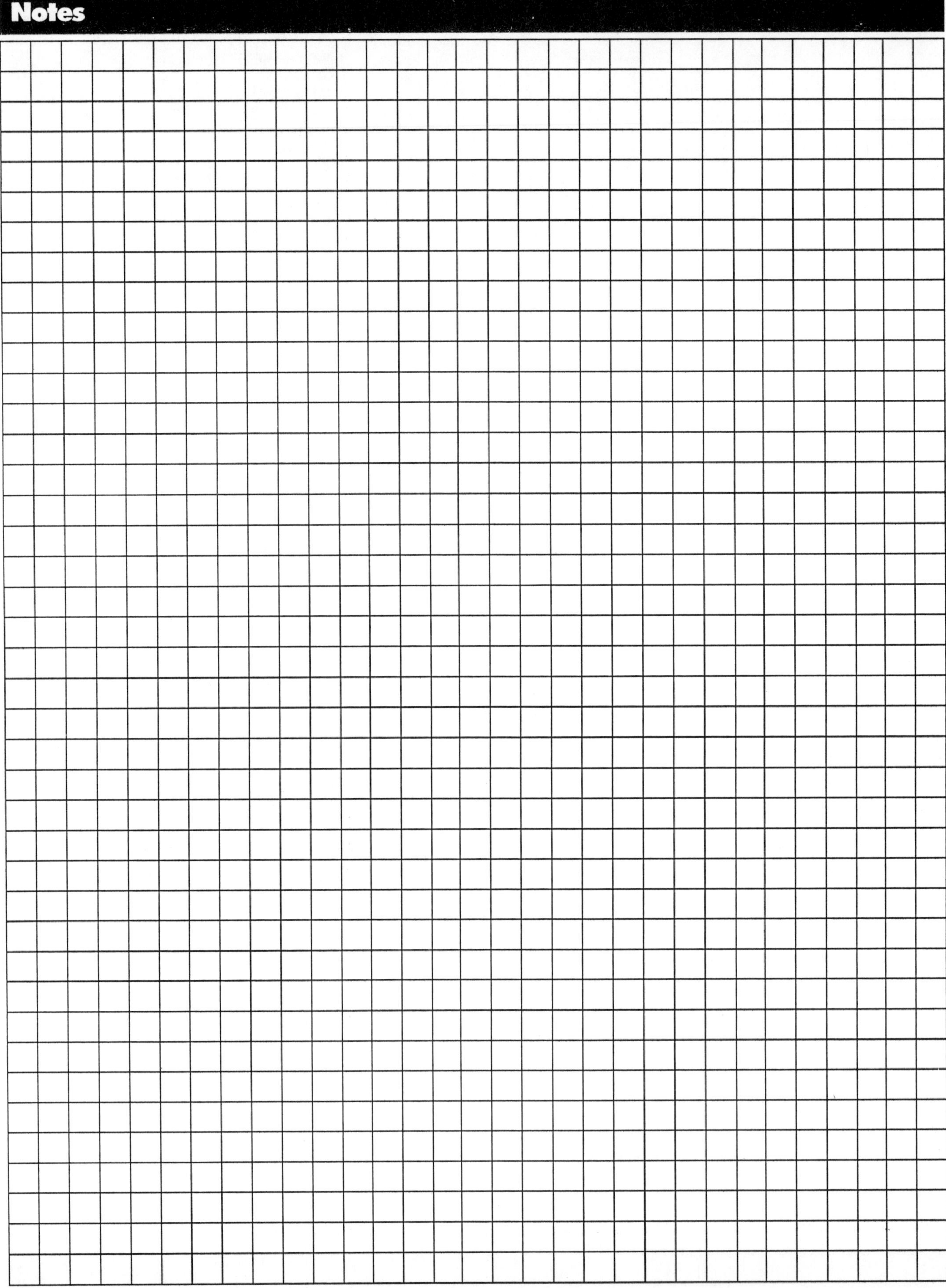

Contractor's Pricing Guides

Means ADA Compliance Pricing Guide

Accurately plan and budget for the ADA modifications you are most likely to need... with the first available cost guide for business owners, facility managers, and all who are involved in building modifications to comply with the Americans With Disabilities Act.

75 major projects—the most frequently needed modifications—include complete estimates with itemized materials and labor, plus contractor's total fees. Over 260 project variations fit almost any site conditions or budget constraints. Location Factors to adjust costs for 927 cities and towns.

A collaboration between Adaptive Environments Center, Inc. and R.S. Means Engineering Staff.

$69.95 per copy
Over 300 pages, illustrated, softcover
Catalog No. 67310   ISBN 0-87629-351-8

Contractor's Pricing Guide:
Residential Square Foot Costs 1996

Now available in one concise volume, all you need to know to plan and budget the cost of new homes. If you are looking for a quick reference, the model home section contains costs for over 250 different sizes and types of residences, with hundreds of easily applied modifications. If you need even more detail, the Assemblies Section lets you build your own costs or modify the model costs further. Hundreds of graphics are provided, along with forms and procedures to help you get it right.

$39.95 per copy
Over 250 pages, illustrated, 8-1/2 x 11
Catalog No. 60326   ISBN 0-87629-413-1

Contractor's Pricing Guide:
Residential Detailed Costs 1996

Every aspect of residential construction from overhead costs to residential lighting and wiring are all here. All the detail you need to accurately estimate the costs of your work with or without markups—labor-hours required, typical crews and equipment are included as well. When you need a detailed estimate, this publication has all the costs to help you come up with a complete, on the money, price you can rely on to win profitable work.

$36.95 per copy
Over 250 pages, with charts and tables, 8-1/2 x 11
Catalog No. 60336   ISBN 0-87629-414-X

Contractor's Pricing Guide:
Framing & Rough Carpentry 1996

The book contains prices for all aspects of framing and rough carpentry for residential and light commercial construction. Based on information provided by suppliers and wholesalers from around the country. To estimate a job, you simply match the specs to our system, calculate the price extensions, add up the total and adjust to your location with the easy-to-use location factors. It's as simple as that. All the forms you will need to prepare your estimate and present it professionally are included as well. You will also find graphics, charts, tables and checklists to help you work through the estimating process quickly and accurately.

$34.95 per copy
Over 200 pages, illustrated, 8-1/2 x 11
Catalog No. 60316   ISBN 0-87629-412-3

Builders' Costs for 100 Best-Selling Home Plans

This book is the first and only resource to combine summary cost estimates with popular home plans. It contains 100 one-page house plans from Home Planner's, Inc., a leading provider of stock plans, with facing-page summary estimate sheets by R.S. Means. A section titled "Management Tools for Builders" provides guidance for reading and selecting plans, choosing a building site, procuring permits, and more.

$34.95 per copy
Over 275 pages, softcover, 8-1/2 x 11
Catalog No. 67311   ISBN 0-87629-356-9

Home Builders' Cost & Planning Kit

What's included:

Builders' Costs for 250 Home Plans (CD: ROM)

• 250 home plans from Home Planners, Inc.
• Summary cost reports for each plan.
• Search by square foot, style, # of bedrooms, # of baths, cost to build.
• Sample city costs included.

Builders' Costs for 100 Best-Selling Home Plans

(See ad in opposite column for information on this publication.)

$39.95 per set
Catalog No. 67311CD   ISBN 0-87629-386-0

1996 Cost Guides

Means Building Construction Cost Data 1996

Means Building Construction Cost Data 1996 is complete in every way, offering the user more value for the money with the Professional's Estimating Service. Free Estimating Hotline Service to assist our customer needs when using Means data. Free *Change Notice* publication sent quarterly with the latest construction trends... plus updates on key material prices, unit prices and City Cost Indexes. New "Estimating Tips" Section at the beginning of each Division for more informed estimating. Easier to use Reference Information. City Cost Index expanded to 305 cities in U.S. and Canada.

$79.95 per copy
Over 600 pages, softcover
Catalog No. 60016   ISBN 0-87629-387-9

Means Open Shop Building Construction Cost Data 1996

With this manual, you'll find unit prices reflecting today's open shop construction marketplace. Offering detailed breakdowns by labor-hours, crew and daily output, bare costs for materials, open shop labor and equipment, performing contractor's overhead and profit. Designed for ease of use for contractors, owners, and facility managers. Includes research results of open shop rates conducted in conjunction with the ABC.

$82.95 per copy
Over 600 pages, softcover
Catalog No. 60156   ISBN 0-87629-395-X

Means Plumbing Cost Data 1996

A great companion book for *Means Mechanical Cost Data*, this widely used guide has comprehensive data for piping, fixtures, fire protection and much more. Includes over 13,700 unit prices and assemblies costs. Additional unit prices for plumbing components needed for complete estimates. Also, complete coverage of ADA-related work. An expanded City Cost Index for 305 cities in the U.S. and Canada. Now available, a new estimating Hotline Service to assist you when using Means data.

$83.95 per copy
Over 500 pages, softcover
Catalog No. 60216   ISBN 0-87629-396-8

Means Residential Cost Data 1996

Now you can prepare fast, accurate estimates for any residential project at any level of detail. This must-have guide contains square foot costs for 30 finished residences with construction grades for economy, average, custom and luxury. Also includes over 100 commonly used assemblies to compare costs for various systems. You'll also find over 8,000 unit costs showing bare material, labor and equipment costs with overhead and profit. Includes easy-adjust factors for over 900 zip codes in the U.S. and Canada.

$74.95 per copy
Over 550 pages, softcover
Catalog No. 60176   ISBN 0-87629-390-9

Means Electrical Cost Data 1996

The most reliable source of electrical construction costs in North America. Everything you need, from raceways to the most advanced microprocessor-based fire alarm command systems. Contains more than 13,000 line items and almost 1,700 assemblies, with a newly expanded City Cost Index that covers 305 cities in the U.S. and Canada. *Means Electrical Cost Data* 1996 has two "How to Use" sections—one for beginners and one for experienced users. "Checklists and Pointers" appear throughout and cover every estimating detail.

$83.95 per copy
Over 460 pages, softcover
Catalog No. 60036   ISBN 0-87629-402-6

Means Repair & Remodeling Cost Data 1996

Commercial/Residential

Commercial or residential, if you'll be estimating renovation or repair projects you need *Means Repair and Remodeling Cost Data* 1996. This edition contains unit costs for over 21,000 components, assemblies costs for over 150 systems, plus hundreds of new unit price entries. The Means "R & R" book includes the newly updated City Cost Index that has been expanded to 305 locations in the U.S. and Canada.

$79.95 per copy
Over 600 pages, softcover
Catalog No. 60046   ISBN 0-87629-389-5

Means Site Work & Landscape Cost Data 1996

The only nationally recognized source for costing site work and landscape jobs. The publication, presented in an easy-to-use format, has over 17,000 products/services and 5,000 varieties/sizes of stock. Widely used by site work and landscape designers and contractors, builders, suppliers, owners and maintenance personnel.

$86.95 per copy
Over 580 pages, softcover
Catalog No. 60286   ISBN 0-87629-391-7

Means Light Commercial Cost Data 1996

This hands-on publication has the industry-standard "Means" costs for fast, accurate estimating of hundreds of small- to mid-sized commercial, institutional and industrial structures. You'll find over 150 commonly used systems, all based on 1996 non-union labor costs, as well as more than 10,000 construction components with bare costs for materials, equipment, and open shop labor with overhead and profit included. Easy-to-use Location Factors for over 900 cities in the U.S. and Canada.

$76.95 per copy
Over 600 pages, softcover
Catalog No. 60186   ISBN 0-87629-393-3

Books for Builders

Basics for Builders:
Plan Reading & Material Takeoff

A complete course in reading and interpreting building plans—and performing quantity takeoffs to professional standards.

This new book shows and explains, in clear language and with over 160 illustrations, typical working drawings encountered by contractors in residential and light commercial construction. The author describes not only how all common features are represented, but how to translate that information into a material list. Organized by CSI division, each chapter uses plans, details and tables, and a summary checklist.

$35.95 per copy
Over 420 pages, illustrated, softcover
Catalog No. 67307   ISBN 0-87629-348-8

Basics for Builders:
How To Survive and Prosper in Construction

An easy-to-apply checklist system that helps contractors organize, manage, and market a construction firm successfully. Topics include:

- Identifying profitable markets
- Assessing the firm's capabilities
- Controlled growth
- The bid/no bid decision
- Scheduling
- Job start-up and planning
- Subcontractor management
- Finances and claims

$34.95 per copy
Over 300 pages, illustrated, softcover
Catalog No. 67273   ISBN 0-87629-342-9

Basics for Builders:
How To Make Money in Insurance Repair

Guidance for getting into profitable insurance repair contracting

This book gives you an insider's look at how property insurance companies work with contractors to repair and restore damaged buildings... how adjustors cost out damages, analyze coverages and handle quotations and bids... and how the work is awarded, supervised and paid for.

$36.95 per copy
Over 200 pages, illustrated, softcover
Catalog No. 67267A   ISBN 0-87629-352-6

Basics for Builders:
Framing & Rough Carpentry

A complete, illustrated do-it-yourself course on framing and rough carpentry. The book covers walls, floors, stairs, windows, doors, and roofs, as well as nailing patterns and procedures. Additional sections are devoted to equipment and material handling, standards, codes, and safety requirements.

The "framer-friendly" approach includes easy-to-follow, step-by-step instructions. This practical guide will benefit both the carpenter's apprentice and the experienced carpenter, and sets a uniform standard for framing crews.

$24.95 per copy
Over 125 pages, illustrated, softcover
Catalog No. 67298   ISBN 0-87629-251-1

Interior Home Improvement Costs

Estimates for over 60 interior projects, including:

- Attic/Basement Conversions
- Kitchen/Bath Remodeling
- Stairs, Doors
- Walls/Ceilings

$19.95 per copy
Over 200 pages, illustrated, softcover
Catalog No. 67308   ISBN 0-87629-349-6

Exterior Home Improvement Costs

Quick estimates for 67 projects, including:

- Room additions
- Garages
- Walls, Fences
- Landscaping

$19.95 per copy
Over 200 pages, illustrated, softcover
Catalog No. 67309   ISBN 0-87629-350-X

Practical Pricing Guides for Homeowners and Contractors

These reduced-price editions include updated estimates for the most popular home improvement projects. All are illustrated and include material costs, labor hours, and total project costs. *Plus:* descriptions of the work, level of difficulty, and tips on selecting materials and avoiding pitfalls. Location factors allow cost adjustment to over 900 U.S. locations. New features include a section on how to select and work with a contractor, with handy checklists for organizing projects.

Quantity Takeoff for Contractors:
How to Get Accurate Material Counts
by Paul J. Cook

Contractors who are new to material takeoffs or want to be sure they are using the best techniques will find that this book answers all of their questions, with every step illustrated. A chapter is devoted to each major construction element. Each chapter includes: detailed descriptions of takeoff items, illustrations of the components, complete calculations, guidelines for improving accuracy, and samples of finished quantity sheets.

$35.95 per copy
Over 250 pages, illustrated, softcover
Catalog No. 67262   ISBN 0-87629-268-6

Superintending for Contractors:
How to Bring Jobs in On-time, On-budget
by Paul J. Cook

Today's superintendent has become a field project manager, directing and coordinating a large number of subcontractors, and overseeing the administration of contracts, change orders, and purchase orders. This book examines the complex role of the superintendent/field project manager, and provides guidelines for the efficient organization of this job.

$35.95 per copy
Over 220 pages, illustrated, softcover
Catalog No. 67233   ISBN 0-87629-272-4

Bidding for Contractors:
How to Make Bids that Make Money
by Paul J. Cook

In this best-selling book, the author shares the benefits of his more than 30 years of experience in construction project management, providing contractors with the tools they need to develop competitive bids. Cook's methods apply to all job sizes, up to multimillion dollar projects.

$35.95 per copy
Over 225 pages, illustrated, softcover
Catalog No. 67180   ISBN 0-87629-270-8

Estimating for Contractors:
How to Make Estimates that Win Jobs
by Paul J. Cook

This widely used reference offers clear, step-by-step estimating instructions that lead to achieving the following goals: objectivity, thoroughness, and accuracy.

Estimating for Contractors is a reference that will be used over and over, whether to check a specific estimating procedure, or to take a complete course in estimating.

$35.95 per copy
Over 225 pages, illustrated, softcover
Catalog No. 67160   ISBN 0-87629-271-6

Business Management for Contractors:
How to Make Profits in Today's Market
by Paul J. Cook

This book focuses on the management of small-to-medium-sized construction companies. Part I addresses the manager's role in the organization, Part II is concerned with procuring contracts, and Part III deals with the manager's responsibilities in fulfilling the contract.

$35.95 per copy
Over 230 pages, illustrated, softcover
Catalog No. 67250   ISBN 0-87629-269-4

Building Spec Homes Profitably
by Kenneth V. Johnson

The author offers a system to reduce risk and ensure profits in spec home building no matter what the economic climate. Includes:

• The 3 Keys to Success: location, floor plan and value
• Market Research: How to perform an effective analysis
• Site Selection: How to find and purchase the best properties
• Financing: How to select and arrange the best method
• Design Development: Combining value with market appeal
• Scheduling & Supervision: Expert guidance for improving your operation

$29.95 per copy
Over 200 pages, softcover
Catalog No. 67312   ISBN 0-87629-357-7

Means On-Line Product Catalog:

Perfect for 24-hour ordering—find the construction cost and reference information you need and order on-line. Look for our exclusive "On-Line Special Offers".

Useful Construction Cost & Reference Info

Features critical construction cost line items... quarterly cost updates... city cost trends... articles from Means' *Change Notice*... excerpts from Means reference books.

List of Construction-Related Internet Sites

Means' webmaster has spent hundreds of hours surfing the net to find construction-related web sites that you'll definitely want to visit. Save time and start here to explore the Internet.

What's New at R.S. Means

Learn about Means' newest products and services and how they can help you make a difference.

DemoSource On-Line

If you're looking for estimating software, come to DemoSource On-Line. Find the software that meets your needs and download a FREE demo.

e-mail Connection

If you have technical questions, problems or comments, now you can e-mail Means engineers and customer service staff!

How do I reach Means On-Line?

A number of on-line service providers, including America Online, Compuserve, Prodigy and The Microsoft Network, give you access to the Internet's worldwide web. If you have one of these services, you can reach **Means On-line** at:
http://www.rsmeans.com

For those who do not have Internet access:

DemoSource™

One-stop shopping for the latest cost estimating software for just $19.95

DemoSource™ — The evaluation tool for estimating software. Includes product literature and demo diskettes for ten or more estimating systems, all of which link to MeansData™.

CALL 1-800-448-8182 TO ORDER!

Qty.	Book No.	COST ESTIMATING BOOKS	Unit Price	Total
	60066	Assemblies Cost Data 1996	$134.95	
	60016	Building Construction Cost Data 1996		
	61016	Building Const. Cost Data—Looseleaf Ed. 1996	109.95	
	63016	Building Const. Cost Data—Metric Ed. 1996	99.95	
	60226	Building Const. Cost Data—Western Ed. 1996	89.95	
	60116	Concrete & Masonry Cost Data 1996	77.95	
	60146	Construction Cost Indexes 1996	198.00	
	60146A	Construction Cost Index—January 1996	49.50	
	60146B	Construction Cost Index—April 1996	49.50	
	60146C	Construction Cost Index—July 1996	49.50	
	60146D	Construction Cost Index—October 1996	49.50	
	60316	Contractor's Pricing Guide: Framing & Carpentry	34.95	
	60336	Contractor's Pricing Guide: Resid. Detailed Costs	36.95	
	60326	Contractor's Pricing Guide: Resid. S.F. Costs	39.95	
	60236	Electrical Change Order Cost Data 1996	89.95	
	60036	Electrical Cost Data 1996	83.95	
	60206	Facilities Construction Cost Data 1996	199.95	
	50300	Facilities Maintenance & Repair Cost Data 1996	199.95	
	50300R	Facilities Maintenance & Repair Renewal	119.95	
	60166	Heavy Construction Cost Data 1996	86.95	
	60096	Interior Cost Data 1996	83.95	
	60126	Labor Rates for the Const. Industry 1996	184.95	
	60186	Light Commercial Cost Data 1996	76.95	
	60026	Mechanical Cost Data 1996	83.95	
	60156	Open Shop Building Const. Cost Data 1996	82.95	
	60216	Plumbing Cost Data 1996	83.95	
	60046	Repair and Remodeling Cost Data 1996	79.95	
	60176	Residential Cost Data 1996	74.95	
	60286	Site Work & Landscape Cost Data 1996	86.95	
	60056	Square Foot Costs 1996	95.95	
		REFERENCE BOOKS		
	67147A	ADA in Practice	69.95	
	67310	ADA Pricing Guide	69.95	
	67305	Asia Pacific Constr. Costs Handbook	139.95	
	67275	Avoiding and Resolving Construction Claims	62.95	
	67298	Basics for Builders: Framing & Rough Carpentry	24.95	
	67273	Basics for Builders: How to Survive and Prosper	34.95	
	67267A	Basics for Builders: Insurance Repair	36.95	
	67307	Basics for Builders: Plan Reading & Takeoff	35.95	
	67180	Bidding for Contractors	35.95	
	67311	Builders' Costs for 100 Home Plans	34.95	
	67261	Building Profess. Guide to Contract Documents	62.95	
	67312	Building Spec Homes Profitably	29.95	
	67250	Business Management for Contractors	35.95	
	67146	Concrete Repair & Maintenance Illustrated	64.95	
	67278	Construction Delays	67.95	
	67268	Construction Paperwork	49.95	
	67255	Contractor's Business Handbook	42.95	
	67230A	Electrical Estimating Methods—2nd Ed.	59.95	
	67160	Estimating for Contractors	35.95	
	67276	Estimating Handbook	99.95	
	67300	European Construction Costs Handbook	139.95	
	67249	Facilities Maintenance Management	84.95	
	67246	Facilities Maintenance Standards	159.95	
	67264	Facilities Manager's Reference	84.95	
	67301	Facilities Planning & Relocation	109.95	

Qty.	Book No.	REFERENCE BOOKS (Cont'd)	Unit Price	Total
	67286	Fire Protection: Design Criteria, Options, Select.	$ 74.95	
	67231	Forms for Building Construction Professionals	94.95	
	67288	Forms for Contractors	79.95	
	67260	Fundamentals of the Construction Process	69.95	
	67210	Graphic Construction Standards	124.95	
	67258	Hazardous Material & Hazardous Waste	89.95	
	67148	Heavy Construction Handbook	72.95	
	67308	Home Improvement Costs—Interior Projects	19.95	
	67309	Home Improvement Costs—Exterior Projects	19.95	
	67304	How to Estimate with Metric Units	49.95	
	67306	HVAC: Design Criteria, Options, Select.-2nd Ed.	84.95	
	67281	HVAC Systems Evaluation	72.95	
	67282	Illustrated Construction Dictionary, Condensed	59.95	
	67292	Illustrated Construction Dictionary, Unabridged	99.95	
	67237	Interior Estimating	62.95	
	67295	Landscape Estimating-2nd Ed.	62.95	
	67266	Legal Reference for Design & Construction	109.95	
	67299	Maintenance Management Audit	64.95	
	67299A	Audit Disk	19.95	
	67302	Managing Construction Purchasing	62.95	
	67294	Mechanical Estimating-2nd Ed.	62.95	
	67245	Planning and Managing Interior Projects	74.95	
	67283	Plumbing Estimating	59.95	
	67236A	Productivity Standards for Constr.-3rd Ed.	159.95	
	67247	Project Planning and Control for Construction	62.95	
	67262	Quantity Takeoff for Contractors	35.95	
	67265	Repair & Remodeling Estimating	69.95	
	67253	Roofing: Design Criteria, Options, Selection	62.95	
	67291	Scheduling Manual-3rd Ed.	62.95	
	67145	Square Foot Estimating	67.95	
	67241	Structural Steel Estimating	79.95	
	67287	Successful Estimating Methods	64.95	
	67287A	Successful Estimating Disk	N/C	
	67313	Successful Interior Projects	69.95	
	67233	Superintending for Contractors	35.95	
	67284	Understanding Building Automation Systems	74.95	
	67259	Understanding Legal Aspects of Design/Build	74.95	
	67303	Unit Price Estimating Methods-2nd Ed.	59.95	

MA residents add 5% state sales tax

Shipping & Handling**

Total (U.S. Funds)*

Mail to:
R.S. Means Company, Inc.
100 Construction Plaza
P.O. Box 800
Kingston, MA 02364-0800

Prices are subject to change and are for U.S. delivery only. *Canadian customers may call for current prices. **Shipping & handling charges: Add 6.5% of total order for check and credit card payments. Add 9% of total order for invoiced orders.

BBAD-1000

Send Order To:

Name (Please Print) ___________________

Company ___________________

☐ Company
☐ Home   Address ___________________

____

City/State/Zip ___________________

Phone # ___________ P.O. # ___________
(Must accompany all orders being billed)